T0230504

Lecture Notes in Electrical Engineering

Volume 441

About this Series

"Lecture Notes in Electrical Engineering (LNEE)" is a book series which reports the latest research and developments in Electrical Engineering, namely:

- Communication, Networks, and Information Theory
- Computer Engineering
- Signal, Image, Speech and Information Processing
- Circuits and Systems
- Bioengineering

LNEE publishes authored monographs and contributed volumes which present cutting edge research information as well as new perspectives on classical fields, while maintaining Springer's high standards of academic excellence. Also considered for publication are lecture materials, proceedings, and other related materials of exceptionally high quality and interest. The subject matter should be original and timely, reporting the latest research and developments in all areas of electrical engineering.

The audience for the books in LNEE consists of advanced level students, researchers, and industry professionals working at the forefront of their fields. Much like Springer's other Lecture Notes series, LNEE will be distributed through Springer's print and electronic publishing channels.

More information about this series at http://www.springer.com/series/7818

Mauro Parodi · Marco Storace

Linear and Nonlinear Circuits: Basic & Advanced Concepts

Volume 1

 Springer

Mauro Parodi
Department of Electrical, Electronic,
  Telecommunications Engineering
  and Naval Architecture (DITEN)
University of Genoa
Genoa
Italy

Marco Storace
Department of Electrical, Electronic,
  Telecommunications Engineering
  and Naval Architecture (DITEN)
University of Genoa
Genoa
Italy

ISSN 1876-1100   ISSN 1876-1119   (electronic)
Lecture Notes in Electrical Engineering
ISBN 978-3-319-87030-4   ISBN 978-3-319-61234-8   (eBook)
DOI 10.1007/978-3-319-61234-8

Printed on acid-free paper

This Springer imprint is published by Springer Nature
The registered company is Springer International Publishing AG
The registered company address is: Gewerbestrasse 11, 6330 Cham, Switzerland

To our past, present, and the future students.

Foreword

Circuit Theory is the bedrock of Electrical and Electronic Engineering. Traditionally, the didactic treatment of the subject has been restricted to linear framework. While linear methods are extremely powerful and universally applied in Engineering practice, they miss the richness of nonlinear phenomena.

This book takes a unified approach to both Linear and Nonlinear Circuits, emphasizing from the outset, that Circuit Theory provides a powerful paradigm for modelling and analysing a range of physical systems, not just the electrical and electronic ones. The laws of Circuit Theory are rooted in fundamental physical principles, and the methods rely on profound mathematics. The circuits paradigm provide accessible insight and intuition into the operation of a wide range of physical systems, from neural networks to coupled pendula.

Parodi and Storace have done an excellent job in presenting foundational material that every undergraduate student should know, followed by advanced concepts which will enrich the learning experience for more advanced undergraduates and graduate students. By combining this material, their book will also be an invaluable resource for practising researchers who will, for the first time, find both classical linear circuits concepts and advanced results in the one place.

Michael Peter Kennedy

Preface

Circuit theory is a discipline at the crossroad between physics, mathematics and system theory and provides basic knowledge in various fields of engineering, such as electronic devices and circuits, signal processing, control systems. The circuit theory core is the concept of *model*, i.e., a mathematical description of a physical system. Circuit theory treats circuits with a field-less approach and, contrary to general electromagnetic theory, neglects the electromagnetic propagation.

Circuit theory was developed later than electrical circuits, whose origin (neglecting the Greek discover of static electricity) dates back to the end of XVIII century. The first circuit theory laws were provided in the middle of XIX century and, up to the middle of XX century, the development of circuit theory was basically related to electrical circuits. During World War II, circuit theory received a strong boost due to the influence and growth of disciplines such as system theory, automatic controls and electronics. With the development of transistors and computers, the realization criteria and the potentialities of circuits underwent deep changes, concerned with both design (the semiconductor miniaturization made it possible to conceive ever more ambitious, complex and low-cost circuits) and the growing corpus of knowledge that circuit theory was called to provide and to rationalize. The devices made available by technology were described by models, first of elementary devices, such as transistors, and subsequently of circuit parts to be used as *building blocks* in broader architectures. The detailed behavior of the devices inside these blocks became unessential for the user and, as such, usually ignored.

Over the years, the need of both designing increasingly complex circuits (largely containing nonlinear devices) and checking their behaviors before concretely realize them required more and more the use of computer numerical simulations. In this perspective, the use of models is essential to provide to the computer a *systematic description* of the circuit structure, together with the electrical properties of its components; furthermore, models contributed to the formalization of *methods* suitable to be mapped into mathematical algorithms, in particular, for the treatment of large-size systems of algebraic equations and of systems of ordinary differential equations. The numerical solutions provided by computers in shorter and shorter times made it possible to study complex circuits also in the absence of an analytical solution.

The circuit theory development had repercussions also outside the circuit field, in areas seemingly far from circuits, but where experimental observations displayed a strong similarity to physical behaviors observable in circuits. This led to formulate equivalent circuit models mimicking the observed phenomena. When World War II was still in full swing, for example, Hodgkin and Huxley formulated their circuit model for neurons, now head of a family of models.

In our vision, a circuit theory course must systematically show ideas and methods suitable to study both linear and nonlinear circuits, either passive or active. The presentation of concepts and fundamental methods of analysis must aim not only to allow their application in today's problems, but also to highlight their meaning and potential as tools to understand and address future developments. These needs were clearly expressed during the '60s of the last century, when Charles Desoer and Ernest Kuh adopted them as guidelines for their treatise "Basic Circuit Theory", which became a classic for university education. In 1969, at University of Genoa, Giuseppe Biorci adopted this text (in the original English version) for his course, and in 1972 he was the Editor of its Italian version. Among the university texts written with similar motivations in the following years, it should be definitely mentioned "Linear and Nonlinear Circuits", written by the previous authors and by Leon Chua about twenty years later. More recent books focused on circuits have downsized, in our opinion, this character. Even if the university programs strongly changed in the last decades, giving in some cases a larger space to information at the expenses of knowledge, there is still need of books providing not only a solid background to all students, but also a broader view to the most motivated among them. We strongly believe that circuit theory is a highly educational discipline, not only for Engineers. Indeed, students learn a quite large set of tools and, when dealing with a specific problem, they have to decide what subset of tools can be (or has to be) used to solve it. In our opinion, this capacity of solving non-trivial problems should remain one of the main elements of the scientific cultural baggage. It can certainly be aided and made more powerful by many practical skills, but cannot be replaced by them.

With these guidelines in mind, this book is structured in *multiple reading levels*: each part is split into two chapters (basic level and advanced level), with two independent levels of reading. Moreover, in the advanced level some suggested shortcuts provide simplified reading paths, left to the reader's choice. The basic chapters are aimed at newcomers to circuit theory, especially students taking a first course in the subject. Their organization is largely based on the experience ripened through a one-semester course, taught for several years, first at the Polytechnic University of Milan and then at the University of Genoa. Our goal with these parts is to explain the mathematics needed to handle circuits as clearly and simply as possible, and to show how it can be used to analyze/understand how a circuit works, also through many worked examples. A peculiar feature of this book is its emphasis on examples, showing how the proposed methods can be applied. The advanced chapters are aimed at both basic-level university students driven by curiosity—in the belief that curiosity is the main driver for studying and learning—and higher-level students, up to Ph.D. students and young researchers that want to find a complete treatment of many mathematical aspects related to circuits. The theory is developed systematically, starting with the simplest circuits (linear, time-invariant and resistive) and providing food for thought on nonlinear circuits, potential functions, linear algebra, geometrical interpretations of some results. This is the subject of this volume. Circuits characterized by time-evolution/dynamics will be treated with the same spirit in a second volume. In our opinion, the multiple reading levels help teachers to adapt their course to best meet students' needs and background. In all cases, students should be assigned homework from the problems at the end of each chapter. They could also do computer projects based on circuit simulators and/or build real circuits during lab activities, e.g., to reproduce and check the results of some of the proposed examples. These aspects are left to the teacher experience and availability.

We are indebted to our friend Lorenzo Repetto, who carefully revised the preliminary version of this book reporting bugs and providing detailed comments, many of which were included in the present volume. We also want to remember our colleague and friend Amedeo Premoli, who passed away in 2014 and who shared our vision of circuit theory. The structure of this book reflects at least in part many discussions had in the past with him.

Genoa, Italy   Mauro Parodi
February 2017   Marco Storace

Contents

Part I Circuit Variables and Topology

1 Basic Concepts 3
  1.1 Physical Systems, Models, Circuits 3
  1.2 Components 6
  1.3 Descriptive Variables 6
    1.3.1 Descriptive Variables for Two-Terminal Elements 8
  1.4 Electric Power and Energy 9
  1.5 Kirchhoff's Laws 11
    1.5.1 Loop and Kirchhoff's Voltage Law 12
    1.5.2 Cut-Set and Kirchhoff's Current Law 14
  1.6 Descriptive Variables for n-Terminal Elements 16
    1.6.1 Descriptive Currents 16
    1.6.2 Descriptive Voltages 17
    1.6.3 Descriptive Variables in the Standard Choice
        (Associated Reference Directions) 19
  1.7 Electric Power in n-Terminal Elements 19
  1.8 Problems 21
  Reference 22

2 Advanced Concepts 23
  2.1 Basic Elements of Graph Theory 23
    2.1.1 Graphs of Components and Circuits 26
    2.1.2 Subgraph, Path, Loop, and Cut-Set 27
    2.1.3 Tree and Cotree 29
  2.2 Matrix Formulation of Kirchhoff's Laws 30
    2.2.1 Fundamental Cut-Set Matrix 31
    2.2.2 Fundamental Loop Matrix 34
    2.2.3 Some General Concepts on Vector Spaces
        and Matrices 36
    2.2.4 The Cut-Set and Loop Matrices and Their Associate
        Space Vectors 39

2.3   Tellegen's Theorem   43
2.4   Problems   44
References   45

**Part II   Memoryless Multi-terminals: Descriptive Equations
           and Properties**

3   Basic Concepts   49
3.1   Solving a Circuit: Descriptive Versus Topological Equations   49
3.2   Descriptive Equations of Some Components   51
      3.2.1   Resistor   51
      3.2.2   Ideal Voltage Source   52
      3.2.3   Ideal Current Source   53
      3.2.4   Elementary Circuits   54
      3.2.5   Diode   58
      3.2.6   Bipolar Junction Transistor   63
3.3   General Component Properties   64
      3.3.1   Linearity   64
      3.3.2   Time-Invariance   64
      3.3.3   Memory   65
      3.3.4   Basis   66
      3.3.5   Energetic Behavior   68
      3.3.6   Reciprocity   69
3.4   Thévenin and Norton Equivalent Representations
      of Two-Terminal Resistive Components   73
      3.4.1   Thévenin Equivalent   73
      3.4.2   Norton Equivalent   76
      3.4.3   Comparisons Between the Two Equivalent Models   79
3.5   Series and Parallel Connections of Two-Terminals   80
      3.5.1   Series Connection   80
      3.5.2   Parallel Connection   82
      3.5.3   Numerical Aspects   87
3.6   Resistive Voltage and Current Dividers   87
      3.6.1   Resistive Voltage Divider   87
      3.6.2   Resistive Current Divider   88
3.7   Problems   88
References   92

4   Advanced Concepts   93
4.1   Potential Functions   93
4.2   Content and Cocontent for Two-Terminal Memoryless
      Elements   94
      4.2.1   Content Function   94
      4.2.2   Cocontent Function   96
      4.2.3   Legendre Transformation   98

    4.3   Content and Cocontent for Multiterminal Resistive Elements   99
    4.4   Additivity of Potential Functions   101
    4.5   Kirchhoff's Laws and Variational Principles for Potential
          Functions   102
    4.6   A Particular Variational Result: Resistive Circuits
          that Minimize Potential Functions   107
    4.7   Minimum Heat Theorem   109
    References   110

**Part III   Memoryless Multi-ports: Descriptive Equations and
             Properties**

5   Basic Concepts   113
    5.1   Port and *n*-Ports   113
    5.2   Descriptive Equations of Some Two-Ports   115
          5.2.1   Controlled Sources   115
          5.2.2   Nullor   121
    5.3   Matrix-Based Descriptions for Linear, Time-Invariant, and
          Memoryless Two-Ports   125
          5.3.1   Resistance Matrix   126
          5.3.2   Conductance Matrix   127
          5.3.3   Hybrid Matrices   129
          5.3.4   Transmission Matrices   130
          5.3.5   Generalizations to Higher Numbers of Ports   131
    5.4   Reciprocity of Two-Ports   132
          5.4.1   Matrix R   132
          5.4.2   Matrix H   132
          5.4.3   Matrix T   133
          5.4.4   Remarks and Examples   134
    5.5   Symmetry of Two-Ports   136
          5.5.1   Matrix R   136
          5.5.2   Hybrid Matrices   137
          5.5.3   Transmission Matrices   138
    5.6   Directionality of Two-Ports   138
    5.7   Two-Port Ideal Power Transferitors   139
          5.7.1   Ideal Transformer   140
          5.7.2   Gyrator   144
    5.8   Connections of Two-Ports   148
          5.8.1   Cascade Connection   148
          5.8.2   Series-Series Connection   150
          5.8.3   Parallel-Parallel Connection   151
          5.8.4   Other Connections and Case Studies   151
    5.9   Problems   154
    References   163

6   Advanced Concepts 165
  6.1   Tellegen's Theorem: A Tool to Investigate Circuit Properties 165
  6.2   Passivity... 166
  6.3   Reciprocity.. 166
  6.4   Circuit Biasing and Small-Signal Relations 169
  6.5   Local Activity and Amplification.......................... 171
  6.6   Colored Edge Theorem 174
  6.7   Circuits Consisting of Two-Terminal Elements: No-Gain
        Theorems... 178
  References 181

Part IV   Analysis of Memoryless Circuits

7   Basic Concepts 185
  7.1   Nodal Analysis 185
        7.1.1   Pure Nodal Analysis.............................. 185
        7.1.2   Modified Nodal Analysis 189
        7.1.3   Substitution Rule 190
  7.2   Mesh Analysis... 191
        7.2.1   Pure Mesh Analysis 192
        7.2.2   Modified Mesh Analysis........................... 196
        7.2.3   Substitution Rule 198
  7.3   Superposition Principle 199
  7.4   Substitution Principle.................................... 204
  7.5   Practical Rules.. 205
        7.5.1   Millmann's Rule.................................. 205
        7.5.2   Two-Ports with $T - \Pi$ Structure 207
        7.5.3 $T \rightleftarrows \Pi$ Transformations 210
        7.5.4   Lattice (Bridge) Structures 211
        7.5.5   (Resistive) Ladder Structures 214
  7.6   Problems 219

8   Advanced Concepts 227
  8.1   From Mesh to Loop Currents: A Graph-Based Generalization. ... 227
  8.2   Tableau Method for Linear Time-Invariant Resistive Circuits 230
  8.3   Superposition Theorem 232
  8.4   Equivalent Representations of Memoryless Multiports. 233
        8.4.1   Thévenin Equivalent Representation of Memoryless
                Multiports...................................... 234
        8.4.2   Norton Equivalent Representation of Memoryless
                Multiports...................................... 241
        8.4.3   Hybrid Equivalent Representations of Memoryless
                Multiports...................................... 248
  8.5   Summarizing Comments 248

8.6   Problems   250
References   253

Appendix: Synoptic Tables   255

Solutions   259

Index   271

About the Authors

Mauro Parodi was appointed full professor of basic circuit theory by the Engineering Faculty at the University of Genoa, Italy, in 1985. His scientific and teaching activity has been focusing on nonlinear circuits and systems theory, nonlinear modeling, and mathematical methods for the treatment of experimental data. He is currently affiliated with the Department of Electrical, Electronic, Telecommunications Engineering and Naval Architecture at the University of Genoa, where he is teaching Mathematical Methods for Engineers and Applied Mathematical Modeling.

Marco Storace received a Ph.D. degree in electrical engineering from the University of Genoa, Italy, in 1998. He was appointed full professor by the same university in 2011 and is currently affiliated with the Department of Electrical, Electronic, Telecommunications Engineering and Naval Architecture there. He was visiting professor at EPFL, Lausanne, Switzerland, in 1998 and 2002, respectively. His main research focus is on nonlinear circuit theory and applications, with an emphasis on circuit models of nonlinear systems such as systems with hysteresis and biological neurons, methods for piecewise-linear approximation (and resulting circuit synthesis) of nonlinear systems, bifurcation analysis, and nonlinear dynamics. He is currently teaching basic circuit theory, analog and digital filters, and nonlinear dynamics at the University of Genoa. From 2008 to 2009 he served as an Associate Editor of the *IEEE Transactions on Circuits and Systems II*, and is currently a member of the IEEE Technical Committee on Nonlinear Circuits and Systems (TC-NCAS).

Acronyms

AC	Alternating Current
CCCS	Current-controlled current source
CCVS	Current-controlled voltage source
DC	Direct Current
DP	Driving-Point
KCL	Kirchhoff's Current Law
KVL	Kirchhoff's Voltage Law
l.h.s.	Left-hand side
PWL	PieceWise-Linear
r.h.s.	Right-hand side
SI	International System of Units
VCCS	Voltage-controlled current source
VCVS	Voltage-controlled voltage source

Part I
Circuit Variables and Topology

Chapter 1
Basic Concepts

Learning, without thought, is a snare; thought, without learning, is a danger.

Confucius

Abstract After having introduced the concept of model, in the basic level of this chapter we define components, descriptive variables, reference directions, electric power and energy, and Kirchhoff's laws.

1.1 Physical Systems, Models, Circuits

All branches of science and engineering are based on the concept of *model*.

Scientific model: A description of a physical system or phenomenon, that relies on existing and commonly accepted knowledge and whose aim is to make it easier to understand, define, describe, or simulate.

The concept of model is well described in the following words by Primo Levi [1].

What we commonly mean by "understand" coincides with "simplify": without a profound simplification the world around us would be an infinite, undefined tangle that would defy our ability to orient ourselves and decide upon our actions. In short, we are compelled to reduce the knowable to a schema. [...] This *desire* for simplification is justified, but the same does not always apply to simplification itself, which is a working hypothesis, useful as long as it is recognized as such and not mistaken for reality. The greater part of historical and natural phenomena are not simple, or not simple in the way that we would like.

© Springer International Publishing AG 2018
M. Parodi and M. Storace, *Linear and Nonlinear Circuits:
Basic & Advanced Concepts*, Lecture Notes in Electrical Engineering 441,
DOI 10.1007/978-3-319-61234-8_1

Fig. 1.1 Mechanical system

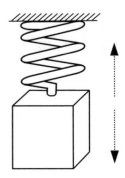

Usually, a model is based on a mathematical description and on a graphical representation and is an approximation of reality. According to Levi's quotation, a simplification implies a working hypothesis, that is, a choice about the elements *necessary and sufficient* to provide the desired level of simplification. This choice should be made by using the so-called Ockham's[1] razor or law of parsimony: among competing hypotheses, the one with the fewest assumptions should be selected.[2]

Often, a model is structured as a set of ideal elements that interact and are, in turn, models. In this book, we call a "circuit" the model of a real circuit and "circuit elements" or "components" the models of real devices.

As stated above, any ideal element is a mathematical/geometrical abstraction aimed to represent or approximate a simple physical event or the behavior of a part of the overall physical system. For instance, the mechanical oscillations of a body vertically suspended to a coil spring and subject to the force of gravity (Fig. 1.1) can be studied by representing the body as a point mass and the spring as a component applying a return force to the body proportional to its elongation.[3] This model of the system ignores the extension and shape of the suspended body, which is considered as a rigid body free from physical extent. Similarly, mass and geometrical dimensions of the coil spring are ignored and the material it is made of is assumed to be perfectly elastic. This body and this spring are the ideal elements of the model of the physical system and are, in turn, models, because the corresponding physical objects (elements) are more complex both in structure and behavior.

Any physical component can be represented by models with different degrees of accuracy. The chosen degree of modeling accuracy usually depends on the complexity

---

[1] William of Ockham (1287–1347) was an outstanding English theologian and scholastic philosopher.

[2] The most popular version of Ockham's principle – "Entities must not be multiplied beyond necessity," translated from Latin Non sunt multiplicanda entia sine necessitate – was actually formulated by the Irish Franciscan philosopher John Punch in his 1639 commentary on the works of Duns Scotus.

[3] This is the so-called Hooke's law, named after the seventeenth-century British physicist Robert Hooke. He first stated the law in 1660 as a Latin anagram and published the solution of his anagram in 1678 as "ut tensio, sic vis," that is, "as the extension, so the force" or "the extension is proportional to the force."

of the system. When the overall complexity exceeds a reasonable level, one usually *changes the scale* of the description. For instance, complex circuits, made up of hundreds or even thousands of elements, are usually represented through global relationships, which neglect the details of what happens to a single device. This is called a behavioral description and aims at describing only the overall *functionality* of the circuit. For instance, integrated circuits are usually modeled in this higher-level way. In other words, the elements modeling parts of a physical system have a mathematical description whose accuracy depends on the scale/complexity of the system itself. Coming back to the previous mechanical example, the simplest model for the spring, for instance, is based on Hooke's law, which ignores the viscous friction originated by the spring movement in the air. A more accurate model would include an ideal element that takes account of this friction. A further refinement could relate to the dissipation of energy due to the imperfect elasticity of the material used for the spring coils, and so on. We thus have the opportunity to represent a system by using physical models where accuracy and complexity grow with those of their components. A model good for accuracy of representation and predictive ability is the basic tool both to study the system's behavior, *analysis*, and to solve *design* problems.

The above general considerations apply, in particular, when the physical system is an electric circuit.

The circuits can be *lumped* – studied by *circuit theory* – or *distributed*. This book deals only with lumped circuits. Their physical dimensions are negligible compared to the smallest wavelength of interest in the electric variables involved. This means that any lumped circuit is made up of lumped components and that the length of any connection is negligible. The lumped circuit, therefore, behaves as a geometric point: any electrical perturbation immediately affects every part of the circuit (null propagation times), so that the spatial position of a component with respect to the others does not affect the electrical behavior of the circuit. In a similar way, in the example of the suspended body to the spring, the body's shape and size were ignored and the body was modeled as a point mass.

In distributed circuits the effects of physical dimensions are no longer negligible. It can be shown, however, that distributed circuits can be thought of as the result of a limit process applied to a suitable sequence of lumped circuits. Therefore a tool for the study of distributed circuits is again circuit theory.

One and the same model can sometimes be used to study physical systems different in nature. For example, it could be shown that the behavior of the mechanical model for the system of Fig. 1.1 is equivalent to that of a circuit made up of two components (namely, an inductor and a capacitor). Generally speaking, this equivalence is based on the correspondences that are established between the physical variables of two models and that allow one to switch from the (mathematical) description of a system to the other. Whenever such equivalence can be established, therefore, an electric circuit can be employed as an equivalent model for a physical system of nonelectrical type.

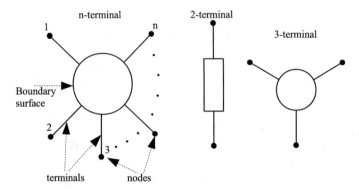

Fig. 1.2 Examples of components

1.2 Components

Any circuit is made up of *components*.

> **Component**: An object delimited by a closed boundary surface from which at least two *terminals* emerge.

The number of terminals is the first element of classification for a component, as shown in Fig. 1.2. The shape and length of any terminal can be adapted to facilitate its connection to other components. The connection between two components is established by the contact of (at least) a terminal edge, or *node*, of each component. It is assumed that all the electromagnetic events characterizing the electrical behavior of the component are confined in the region within the boundary surface and that the overall electrical behavior of the component can be described in terms of variables (called *descriptive variables*) that can be *measured* in the region external to the boundary surface. Therefore the descriptive variables adopt a scale of representation that ignores the detail of electromagnetic events within the boundary surface. This means that in circuit theory any information concerning the physical structure and electromagnetic fields within the boundary surface of the components is not considered. As an analogy, a newspaper article could be identified citing the title and its author, but neglecting the detailed information represented by the words of its text.

1.3 Descriptive Variables

The descriptive variables for a component must be measurable outside its boundary surface. The most common choice is given by *voltages and currents*. Alternatively, one might consider magnetic fluxes and electric charges, even if less easily measurable.

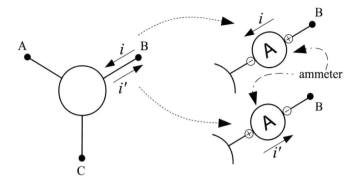

Fig. 1.3 Ammeter connections

Current: A real variable $i \in \mathbb{R}$ which, in the general case, depends on time: $i = i(t)$. It is associated with each terminal of the component according to a given convention. The SI unit of measurement of a current is the *ampere*, whose symbol is A.[4]

The current through a terminal is measured using an instrument called the ammeter. It is assumed that the ammeter is ideal, namely that the application of this instrument does not alter the behavior of the circuit. A conventional arrow defines the way in which the current must be measured. The ammeter is a particular two-terminal element whose nodes (marked as + and −) are represented adjacent to the boundary surface (i.e., the terminals are not represented).

The direction of the arrow representing the current in a terminal defines how to connect the ammeter. In Fig. 1.3 the two possible choices i and i' for the current in the terminal B are indicated by arrows in opposite directions ($i' = -i$). They correspond to two opposite orientations of the pair + and − for the ammeter.

If the ammeter indication in the top connection of Fig. 1.3 is -3 A, it is said that $i = -3$ A; inverting the nodes + and − as in the bottom connection, the ammeter indication becomes $+3$ A and it is said that $i' = +3$ A.

Voltage: A real variable $v \in \mathbb{R}$ which, in the general case, depends on time: $v = v(t)$. It is associated with any pair of nodes in the circuit according to a given convention. The SI unit of measurement of voltage is the *volt*, whose symbol is V.[5]

---

[4]The ampere is named after André-Marie Ampère (1775–1836), a French mathematician and physicist.

[5]The volt is named in honor of the Italian physicist Alessandro Volta (1745–1827), who invented the voltaic pile, possibly the first chemical battery.

Fig. 1.4 Voltmeter connections

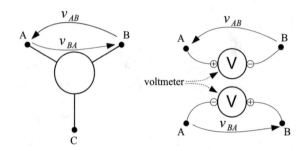

The voltage is measured using an instrument called the voltmeter. It is supposed that the voltmeter is ideal; that is, its connection to any pair of nodes does not modify the behavior of the circuit. An oriented arc between the pair of nodes defines the way in which the voltage must be measured. In Fig. 1.4 the two possible choices v_{AB} and v_{BA} for the voltage between the nodes A and B are indicated by oppositely oriented arcs ($v_{BA} = -v_{AB}$). The corresponding voltmeter connections differ in the opposite orientations of the pair + and −.

If the voltmeter indication in the top connection of Fig. 1.4 is $1.3V$, it is said that $v_{AB} = 1.3V$; inverting the nodes + and - as in the bottom connection, the voltmeter indication becomes $-1.3V$ and it is said that $v_{BA} = -1.3V$.

1.3.1 Descriptive Variables for Two-Terminal Elements

We now consider a two-terminal element, which is the simplest element that can be defined.[6] On the basis of the previous section, for each terminal we can introduce two currents of opposite direction, as shown in Fig. 1.5a.

Actually, in a two-terminal element we always have $i_1 = i_3$ and $i_2 = i_4$. This property, which for now we assume as valid without proof, follows directly from Kirchhoff's current law, discussed in Sect. 1.5. By virtue of this property, the number of actual choices for the current reduces to two, identified as i_1 and i_2 in Fig. 1.5b. Obviously, we have $i_1 = -i_2$. Summing up, the currents and the voltages that can be measured on a two-terminal element are i_1, i_2, v_1, v_2 as shown in Fig. 1.5c. However, in pair (i_1, i_2) only one current needs to be measured, and in pair (v_1, v_2) only one

---

[6]For a one-terminal element no voltage can be defined, because we need at least two nodes. A one-terminal electric structure makes sense only when propagation phenomena are present (e.g., an antenna).

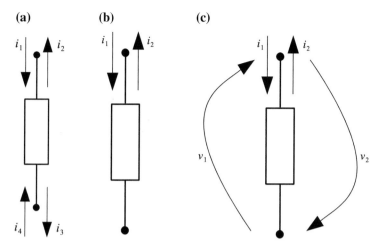

Fig. 1.5 Voltages and currents in a two-terminal element

voltage needs to be measured: the remaining variables are obtained from the known ones through a simple change of sign. Thus the amount of information necessary and sufficient to characterize the two-terminal element is given by the values of one voltage and one current. These two variables are the descriptive variables for the two-terminal element.

We still have one possible choice concerning the relative orientation of current and voltage. Henceforth, the chosen convention or *standard choice* (or *associated reference directions*) is to assume that they have opposite relative orientations (as the (v, i) pair shown in Fig. 1.6a): this is not mandatory, of course, as well as the choice of currents and voltages as descriptive variables. But current and voltage, generally speaking, are time-dependent variables, thus they can change their value and their sign with time. Therefore it is usually meaningless to decide a priori the *actual* reference directions for voltages and currents. The standard choice sets up a reference scenario to measure/determine variables without any ambiguity and allows introducing the component equation and the definition of electric power absorbed/delivered by the component.

1.4 Electric Power and Energy

We can now define a third fundamental variable: electric power.

Absorbed power: For a two-terminal component described by variables taken according to the standard choice, the *absorbed* (electric) power is defined as

$$p(t) = v(t)i(t) \tag{1.1}$$

Fig. 1.6 Standard choice **(a)** and nonstandard choice **(b)** for the reference directions of current and voltage

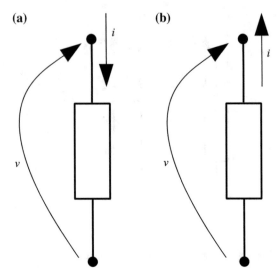

The (derived) SI unit of measurement of power is the *watt*, whose symbol is W.[7]

Notice that, owing to the definition of electric power given above, $1W = 1V \times 1A$. The instrument used to measure the power absorbed by a two-terminal is called a *wattmeter*.

The power *delivered* by the component is the *opposite* (in sign) of the absorbed power.

We remark that these definitions hold only when the variables are taken according to the standard choice; if this is not the case, before computing a power it is necessary to change the sign and orientation to one variable to restore the reference scenario. To clarify this concept, it is useful to resort to a case study.

Case Study

In the example shown in Fig. 1.7a, the current i ($= 2\,mA$) and the voltage v ($= 1.5\,V$) satisfy the standard choice. Then, the absorbed power is $p(t) = v(t)i(t) = 3\,mW$ and the delivered power is $\tilde{p}(t) = -3\,mW$. On the contrary, in the case shown in Fig. 1.7b, the current i' ($= 1\,mA$) and the voltage v ($= 2\,V$) do not satisfy the standard choice. Then, before computing any power, we have to restore the reference scenario, for example, by acting on the current, as shown

---

[7]The watt is named after the Scottish scientist James Watt for his contributions to the development of the steam engine.

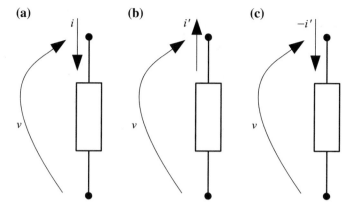

Fig. 1.7 Examples of power computation (See text)

in Fig. 1.7c. Then, the absorbed power is $p(t) = v(t)(-i'(t)) = -2\,\mathrm{mW}$ and the delivered power is $\tilde{p}(t) = 2\,\mathrm{mW}$.

A strictly related variable is the absorbed (electric) energy, defined as

$$w(t) = \int_{-\infty}^{t} p(\tau)d\tau \qquad (1.2)$$

In other terms, $p(t) = \frac{dw}{dt}$.
The energy absorbed/delivered by a component can be

- *Converted/transformed in another form of energy*: For instance, the lamps convert the electric energy into light and (partially) heat, the heaters into heat and (partially) light, the air fan and the food processor into mechanical energy, and so on.
- *Stored*: A device that captures energy produced at one time for use at a later time is sometimes called an accumulator. For instance, a rechargeable battery stores readily convertible chemical energy to operate a mobile phone. In other contexts, a dam stores energy in a reservoir as gravitational potential energy, fossil fuels store ancient energy derived from sunlight, and food is a form of energy stored in chemical form.

1.5 Kirchhoff's Laws

In a circuit made up of N two-terminals we have $2N$ descriptive variables, that is, N voltages and N currents. In order to find the values of all of them, is it necessary to measure all of them? We look for an answer by considering the simple example shown in Fig. 1.8, where voltages have been assigned arbitrarily and currents have

Fig. 1.8 A circuit made up
of $N = 7$ two-terminals

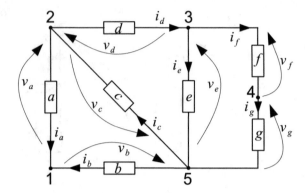

been defined according to the standard choice. This example is used in the following
subsections to introduce the two Kirchhoff's laws,[8] which are based on the concepts
of loop and cut-set.

1.5.1 Loop and Kirchhoff's Voltage Law

The word "circuit" recalls a closed path or trip around something (e.g., one circuit of a
planet around the sun). Strictly speaking, an electric circuit is just a set of components
allowing a closed path of voltages: in this sense it is a synonym of loop. (See below.)
In a wide sense, it can denote more complex structures, containing several possible
paths, and is synonymous with *electric network*.

> **Loop**: A closed path (or surface, more in general) corresponding to a sequence
> (with two possible orientations) of $n - 1$ distinct nodes. The nth node must
> coincide with the first one to close the loop.

Figure 1.9 shows (dashed lines) three examples of loops for the circuit of Fig. 1.8.
    A loop can be oriented either clockwise or counterclockwise, according to a given
criterion or arbitrarily. For instance, Fig. 1.10 shows arbitrary orientations for the
loops of Fig. 1.9. Henceforth, a loop is usually associated with a specific orientation.
Following the definition given in Sect. 1.3, the sequence of nodes involved by a loop
is related to a set of voltages. Each voltage of a given loop can be oriented either like
or unlike the loop.
    If we measure with a set of voltmeters all the voltages contained in a loop and
sum them by assigning a positive sign to the voltages oriented according to the loop

---

[8]Gustav Robert Kirchhoff (1824–1887) was a German physicist who contributed to the fundamental
understanding of electrical circuits, spectroscopy, and the emission of black-body radiation by heated
objects.

Fig. 1.9 Three examples of loops. Sequences: [1,2,5] or [1,5,2] (loop A), [2,3,4,5] or [2,5,4,3] (loop B), [1,2,3,4,5] or [1,5,4,3,2] (loop C)

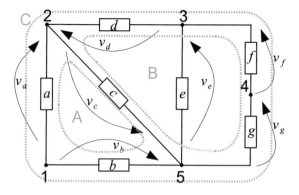

Fig. 1.10 Examples of loop orientations

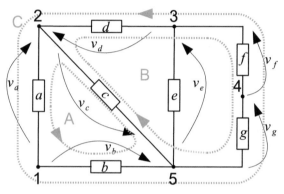

and a negative sign to the voltages with opposite orientation, we see that this sum is identically equal to $0\,V$.

> **Kirchhoff's Voltage Law (KVL)**: In a loop, the sum of the voltages oriented like the loop is equal to the sum of the voltages with opposite orientation.

For instance, loop A in Fig. 1.10 (oriented counterclockwise) involves voltages v_a and v_c with opposite orientation and voltage v_b with the same orientation. Thus the KVL for this loop states that $v_a + v_c = v_b$. This means that we can measure just two of these voltages, inasmuch as the third one is not independent of the other two. For instance, if we measure with a voltmeter $v_a = 3\,V$ and $v_b = 1.5\,V$, we know without need of further measure that $v_c = -1.5\,V$. It is evident that the loop orientation is completely arbitrary, because it does not influence the relationship between the three variables.

Similarly, loop B (oriented clockwise) involves voltages v_d, v_c, v_f, and v_g with the same orientation. The KVL for this loop states that $v_d + v_c + v_f + v_g = 0$. Then, knowing that $v_c = -1.5\,V$, if we measure with a voltmeter $v_f = 1\,V$ and $v_g = 2.5\,V$, the KVL ensures that $v_d = -2\,V$.

Fig. 1.11 Three examples of cut-sets

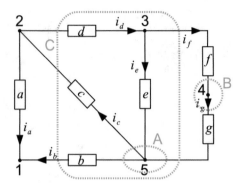

To check your comprehension, you can try to:

- Check the validity of KVL for loop C, with the voltage values listed above.
- Find other loops in the circuit and also check the validity of KVL for them.

For a given circuit, a loop not containing any component inside or outside is called henceforth *mesh*. The (unique) mesh not containing any outside component is henceforth called the *outer loop*, whereas the meshes not containing any inside component are called *inner loops*. In the considered example, mesh C is the outer loop and mesh A is one of the three possible inner loops.

1.5.2 Cut-Set and Kirchhoff's Current Law

Concepts similar to those introduced in the previous section for voltages hold for currents.

> **Cut-set**: A set of n distinct terminals crossed by a closed path. If we imagine that these terminals are cut, the resulting circuit would be made up of two or more distinct parts.

Figure 1.11 shows (dashed lines) three examples of cut-sets for the circuit of Fig. 1.8. For closed paths (henceforth called in turn cut-sets) A and B, one of the two parts the circuit is made up of after cutting the corresponding terminals comprises only one node. (See Fig. 1.12, panels a and b.) For cut-set C, the circuit resulting after cutting the involved terminals is made up of three parts. (See Fig. 1.12c.)

A cut-set can be oriented either inwards or outwards, according to a given criterion or arbitrarily. For instance, Fig. 1.13 shows arbitrary orientations for the cut-sets of Fig. 1.11.

Following the definition given in Sect. 1.3, the terminals involved by a cut-set are related to a set of currents. Each current of a given cut-set can be oriented either like or

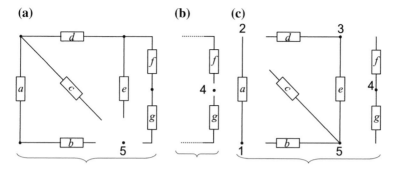

Fig. 1.12 Circuits resulting after removing the cut-sets A (**a**), B (**b**), and C (**c**)

Fig. 1.13 Examples of
cut-set orientations

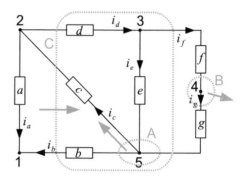

unlike the cut-set. If we measure with a set of ammeters all the currents contained in
a cut-set and sum them by assigning a positive sign to the currents oriented according
to the cut-set and a negative sign to the currents with opposite orientation, we see
that this sum is identically equal to 0 A.

> **Kirchhoff's Current Law (KCL):** In a cut-set, the sum of the currents flowing
> into the cut-set is equal to the sum of the currents with opposite orientation.

For instance, cut-set A in Fig. 1.13 (oriented outwards) cuts terminals of components b-c-e-g and involves currents i_b and i_c with the same orientation (these
currents flow out of the dashed surface) and currents i_e and i_g with opposite orientation (these currents flow into the dashed surface). Then, the KCL for this cut-set
states that $i_b + i_c = i_e + i_g$. This means that we can measure just three of these currents, because the fourth one is not independent of the other three. For instance, if
we measure with an ammeter $i_b = 2\,\text{mA}$, $i_c = -1.5\,\text{mA}$, and $i_e = 1\,\text{mA}$, we know
without need of a further measure that $i_g = -0.5\,\text{mA}$. It is evident that the loop orientation is completely arbitrary, inasmuch as it does not influence the relationship
between the four variables.

Fig. 1.14 KCL for a
two-terminal element

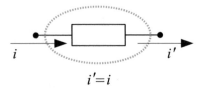

$$i' = i$$

Cut-set B (oriented outwards) simply implies that $i_f = i_g$.

A cut-set containing only one node inside is henceforth called a *node cut-set*. If we cut the corresponding terminals, the resulting circuit is made up of two parts, one of which is the node itself. In the considered example, both A (node 5) and B (node 4) are node cut-sets.

Cut-set C (oriented inwards) cuts terminals of components b-c-d-f-g and involves current i_g and i_d with the same orientation and currents i_b, i_c, and i_f with opposite orientation. The KCL for this cut-set states that $i_g + i_d = i_b + i_c + i_f$. Then, knowing (from above) that $i_b = 2\,\text{mA}$, $i_c = -1.5\,\text{mA}$, $i_f = i_g = -0.5\,\text{mA}$, the KCL ensures that $i_d = 0.5\,\text{mA}$. Cut-set C is not a node cut-set.

To check your comprehension, you can try to find other cut-sets in the circuit and check the validity of KCL for them also.

Remark As anticipated, KCL justifies the validity of the assumption made in Sect. 1.3 for the currents in a two-terminal element. The cut-set in Fig. 1.14 clearly shows that currents i and i' are always identical: $i = i'$.

1.6   Descriptive Variables for n-Terminal Elements

Kirchhoff's laws allow us to establish how many currents and voltages should be measured to characterize an n-terminal element (or n-terminal component, or simply n-terminal). The solution to this problem generalizes the result established for a two-terminal element.

1.6.1   Descriptive Currents

A generic n-terminal element is represented in Fig. 1.15a. Its n nodes are arbitrarily numbered from 0 to $n - 1$. For simplicity, all the currents at the terminals are assumed entering the boundary surface of the component. The index of each current is the number of the corresponding node. In the figure, the boundary surface of the component is enclosed within the dashed gray cut-set. This cut-set concerns all the terminal currents, whereby the KCL implies:

Fig. 1.15 Currents in an n-terminal (**a**) and in a four-terminal (**b**)

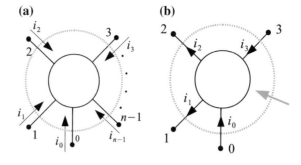

$$\sum_{k=0}^{n-1} i_k = 0 \tag{1.3}$$

and one current can always be expressed in terms of the remaining $n - 1$ currents. In an n-terminal component, therefore, the currents that must be known are $n - 1$. They are referred to as the *descriptive currents* for the component.

As an example, once the currents $i_1, \ldots, i_{n-1}$ are known (say by measurements), the i_0 current is obtained as

$$i_0 = -\sum_{k=1}^{n-1} i_k. \tag{1.4}$$

The conventions adopted above are not binding.

For instance, Fig. 1.15b shows a four-terminal component in which not all the currents are entering the boundary surface. The KCL equation can be written as

$$i_0 - i_1 - i_2 + i_3 = 0 \tag{1.5}$$

and three descriptive currents can be freely selected simply by removing one element from the set $\{i_0, i_1, i_2, i_3\}$.

1.6.2 Descriptive Voltages

The easiest way to identify a set of descriptive voltages is represented in Fig. 1.16a. After selecting a reference node (here denoted by the number 0), we define the voltages $v_1, v_2, \ldots, v_{n-1}$ between this node – said *common* node – and the nodes $1, 2, \ldots, n - 1$. Once these voltages are known, any other voltage between two nodes of the component may be found through KVL.

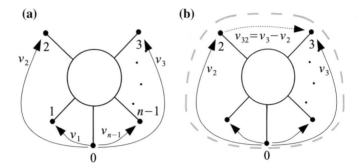

Fig. 1.16 Voltages in an n-terminal element

Fig. 1.17 Voltages in a
five-terminal element. The
four chosen descriptive
voltages are v_a, v_b, v_c, v_d

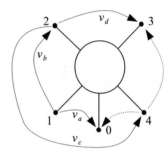

For example, the voltage v_{32} between nodes 3 and 2 (dashed in Fig. 1.16b) is obtained by observing that it forms a loop with the voltages v_3 and v_2. For this loop (dashed gray line) KVL gives $v_2 + v_{32} - v_3 = 0$, so that $v_{32} = v_3 - v_2$. More generally, each voltage v_{AB} between two nodes A and B (both different from the common node) forms a loop with the descriptive voltages v_A and v_B. For this loop, KVL provides $v_{AB} = v_A - v_B$.

The $n - 1$ voltages $v_1, v_2, \ldots, v_{n-1}$ form a set of descriptive voltages for the n-terminal component. They meet the requirements of *independence* (i.e., no pair of descriptive voltages forms a loop) and *completeness* (every other voltage between the nodes of the component can be obtained by means of KVL).

The procedure based on the common node is just one of many to obtain a set of descriptive voltages. Figure 1.17 gives an example for $n = 5$. Here the set was obtained by starting with the voltage v_a between the pair of nodes 0 and 1; the subsequent voltages were obtained by connecting, each time, a new node to one already considered. In this way v_b, v_c, v_d are attained in sequence.

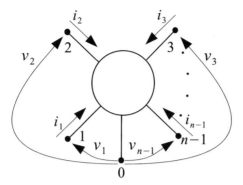

Fig. 1.18 The standard choice for the descriptive variables in an n-terminal component

Whatever the method (including that based on the common node), the number of descriptive voltages is $n - 1$; the first voltage connects the first pair of chosen nodes, and for each of the following only one new node is added.

1.6.3 Descriptive Variables in the Standard Choice (Associated Reference Directions)

The standard choice for the descriptive variables in an n-terminal component easily follows from the foregoing considerations. Obviously, it generalizes the corresponding choice given for a two-terminal component.

The $n - 1$ descriptive voltages are chosen referring to a common node. The corresponding descriptive currents, all entering the component's boundary surface, are obtained by simply discarding the current at the common terminal, as shown in Fig. 1.18.

As a final comment, because the common node can be chosen in n different ways, the number of different standard choices is also n.

1.7 Electric Power in n-Terminal Elements

The use of standard choice for descriptive variables allows the direct calculation of the power absorbed by an n-terminal component.

To show this, consider, for example, the three-terminal component T shown in Fig. 1.19a together with its descriptive variables. The circuit shown in Fig. 1.19b is obtained by connecting T to the two-terminals A and B. In the figure, all the descriptive variables are represented.

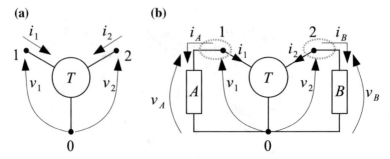

Fig. 1.19 Three-terminal with its descriptive variables (a) and circuit containing the same three-terminal (b)

Any lumped circuit as a whole is an isolated physical system; that is, it cannot exchange power with the outside. This means that the sum of the powers absorbed by all of its components must be zero. In the present case, then, we can write:

$$p + v_A i_A + v_B i_B = 0 \tag{1.6}$$

where p denotes the power absorbed by the three-terminal component T and the remaining terms indicate the power absorbed by A and B, respectively. In the circuit, the KCLs for the gray cut-sets 1 and 2 imply $i_A + i_1 = 0$ and $i_B + i_2 = 0$. Moreover, we have $v_A = v_1$ and $v_B = v_2$. Taking this into account, from Eq. 1.6 we obtain

$$p = \underbrace{v_A(-i_A) + v_B(-i_B)}_{delivered\ by\ A\ and\ B} = v_1 i_1 + v_2 i_2 \tag{1.7}$$

Therefore, the power absorbed by the three-terminal component T, which must equal the sum of the powers delivered by A and B, is simply obtained as the sum of the $v_k i_k$ ($k = 1, 2$) products of the descriptive variables defined according to the standard choice.

The extension of this result to the case of a component with any number n of terminals is entirely obvious.

In terms of standard-choice descriptive variables (Fig. 1.18), the power absorbed by an n-terminal is:

$$p(t) = \sum_{k=1}^{n-1} v_k i_k \tag{1.8}$$

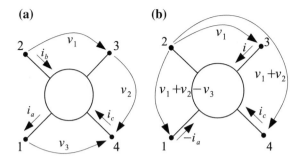

Fig. 1.20 Case Study. The term i is used as an abbreviation for $i_a - i_b - i_c$

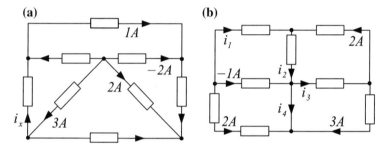

Fig. 1.21 Problems 1.1 (**a**) and 1.2 (**b**)

Case Study

Determine the power absorbed by the four-terminal shown in Fig. 1.20a and express the result in terms of the shown descriptive variables, by ordering with respect to voltages v_1, v_2, v_3.

The first step is to choose a common node, for example, node 2. Then, we express the new descriptive variables (taken according to the standard choice) in terms of the original descriptive variables, as shown in Fig. 1.20b. Finally, we apply Eq. 1.8 and order the resulting expression as requested: $p(t) = v_1(i_a - i_b - i_c) + (v_1 + v_2)i_c + (v_1 + v_2 - v_3)(-i_a) = -v_1 i_b + v_2(i_c - i_a) + v_3 i_a$.

1.8 Problems

1.1 Find the unknown current i_x in the circuit shown in Fig. 1.21a.

1.2 Find all the unknown currents in the circuit shown in Fig. 1.21b.

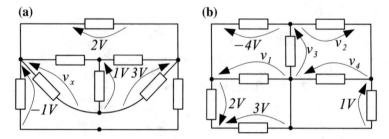

Fig. 1.22 Problems 1.3 (**a**) and 1.4 (**b**)

Fig. 1.23 Problem 1.5

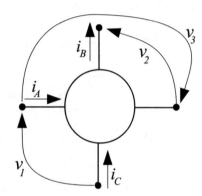

1.3 Find the unknown voltage v_x in the circuit shown in Fig. 1.22a.

1.4 Find all the unknown voltages in the circuit shown in Fig. 1.22b.

1.5 Determine the power absorbed by the four-terminal shown in Fig. 1.23 in terms of the descriptive variables in the figure, by ordering with respect to currents i_A, i_B, i_C.

Reference

1. Levi P (1989) The drowned and the saved. Vintage Books, New York

Chapter 2
Advanced Concepts

Abstract In this chapter we introduce elements of graph theory, graphs of components, matrix formulation of Kirchhoff's laws, matrix associated spaces, and Tellegen's theorem.

2.1 Basic Elements of Graph Theory

A set of independent Kirchhoff's laws for a given circuit – the so-called *topological equations* – can be automatically found by relying on some concepts of *graph theory*, that is, the study of mathematical/geometrical structures, called *graphs*, used to model pairwise relations between objects. Graph theory almost certainly began when, in 1735, Leonhard Euler[1] solved a popular puzzle about the bridges of the East Prussian city of Königsberg (now Kaliningrad) [1]. Nowadays, graph theory is largely used in mathematics, computer science, and network science, but it can be applied in any context where many units interact in some way, such as the components in a circuit. Usually, a graph completely neglects the nature of each unit and of the interactions, just keeping information about their existence.

> A **graph** is a finite set of N *nodes* (or vertices or points), together with a set of L *edges* (or branches or arcs or lines), each of them connecting a pair of distinct nodes.

We remark that more than one edge can connect the same pair of nodes. In this case, these edges are said to be *in parallel*. This implies that a pair of nodes can be insufficient to identify an edge univocally. Moreover, the above definition excludes the degenerate case of edges connecting one node to itself. Henceforth, we label the nodes with numbers and the edges with letters/symbols.

---

[1]Leonhard Euler (1707–1783) was a Swiss mathematician, physicist, astronomer, logician, and engineer who made important and influential discoveries in many branches of mathematics.

© Springer International Publishing AG 2018   23
M. Parodi and M. Storace, *Linear and Nonlinear Circuits:*
Basic & Advanced Concepts, Lecture Notes in Electrical Engineering 441,
DOI 10.1007/978-3-319-61234-8_2

Fig. 2.1 Example of
undirected (**a**) and directed
(**b**) graph with $N = 5$ and
$L = 7$

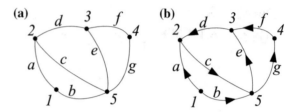

In the simplest case, the edges are not oriented: in this case we have an *undirected graph*, an example of which is shown in Fig. 2.1a. If the edges are oriented, they are called arrows (or directed edges or directed arcs or directed lines) and we have a *directed graph* or *digraph*. (See Fig. 2.1b.)

> **Order** of a node: Number of edges connecting this node to other nodes.

For instance, in the figure node 1 has order 2, node 3 order 3, and node 5 order 4. The specific shape of a graph is not relevant, according to the following definition.

> Two (directed) graphs G_1 and G_2 are **isomorphic** if it is possible to establish a bijective correspondence between:
>
> - Each node of G_1 and each node of G_2
> - Each edge of G_1 and each edge of G_2
>
> such that corresponding edges connect (ordered) pairs of corresponding nodes.

Three examples of graphs isomorphic to the one of Fig. 2.1b are shown in Fig. 2.2. For ease of comparison, we used the same labels for nodes and edges as in Fig. 2.1b; in this case, the graph is not only isomorphic, but is essentially the same. Any change in the labels would not affect the equivalences. The graphs shown in Fig. 2.3 are in turn isomorphic to the one of Fig. 2.1b. Some of the correspondences are summarized in Table 2.1. You can check your comprehension by finding the missing correspondences.

> **Planar graph**: A graph that can be embedded in the plane; that is, it can be drawn on the plane in such a way that all its edges intersect only at their endpoints.

In other words, any planar graph admits an isomorphic graph where no edges cross each other. Some examples of planar graphs are shown in Fig. 2.4.

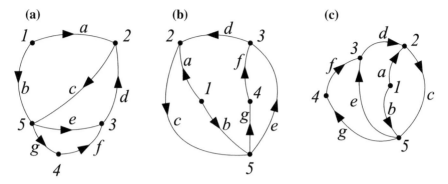

Fig. 2.2 Examples of isomorphic graphs (to be compared to Fig. 2.1b)

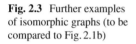**Fig. 2.3** Further examples of isomorphic graphs (to be compared to Fig. 2.1b)

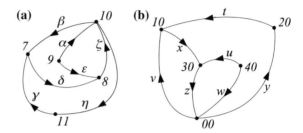

Table 2.1 Table of correspondences between elements of the isomorphic graphs of Figs. 2.1b, 2.3a, and 2.3b

	Graph		
	Fig. 2.1b	Fig. 2.3a	Fig. 2.3b
Graph element	a	ε	u
	b	α	w
	c	ζ	z
	d	δ	x
	1	9	40
	2	8	30
	3	7	10
	4	11	20
	5	10	00

Star graph: A graph containing $N - 1$ nodes of order 1 and one node of order $N - 1$.

Figure 2.5 shows an example of a star graph with 5 nodes: node 5 has order 4; the other nodes have order 1.

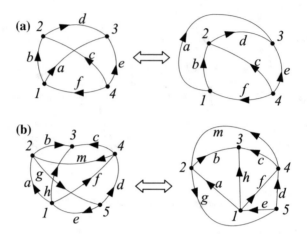

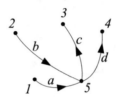

Fig. 2.4 Examples of planar graphs

Fig. 2.5 Example of a star graph

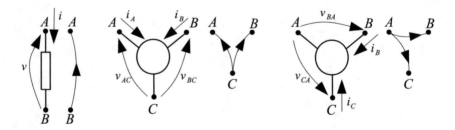

Fig. 2.6 Examples of graphs for multiterminal components

2.1.1   Graphs of Components and Circuits

For a circuit, it is quite natural (even if this is not the only possible choice) to associate the circuit nodes with graph nodes and the descriptive voltages with graph arrows. By assuming the standard choice, this means that each graph arrow is associated with a voltage oriented like the arrow and to a current oriented in the opposite direction.

Figure 2.6 shows some examples of graphs for multiterminal components.

Fig. 2.7 Kuratowski graphs

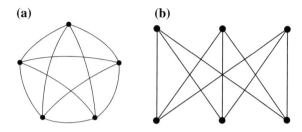

By substituting each circuit component with its graph, we obtain the circuit graph. For instance, the directed graph shown in Fig. 2.1b corresponds to the circuit of Fig. 1.8.

2.1.2 Subgraph, Path, Loop, and Cut-Set

In this section we define some basic graph structures.

> **Subgraph**: A subset of the elements of a given graph, obtained by removing some edges and/or some nodes together with the corresponding edges.

A subgraph is in turn a graph. For instance, by removing edges a, d, f from the graph of Fig. 2.1a, we obtain a subgraph which is in turn a star graph.

It has been shown [2] that a graph is nonplanar if and only if it is (or contains a subgraph) a graph isomorphic to the ones shown in Fig. 2.7, independently of the edge orientations.

> **Path**: A subgraph made up of a sequence of $k-1$ adjacent edges (the orientation is not relevant) connecting a sequence of k nodes that, by most definitions, are all distinct from one another.

In other words, a nondegenerate path is a trail in which all nodes and all edges are distinct and then we have 2 nodes of order 1 (the first and the last) and $k-2$ nodes of order 2. Figure 2.8 shows some examples of paths (in grey) for the reference graph of Fig. 2.1b.

> A graph is **connected** when there is a path between every pair of nodes. Otherwise it is **disconnected**.

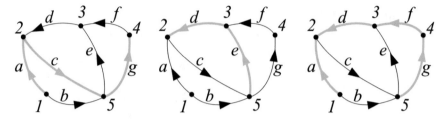

Fig. 2.8 Examples of paths (in *grey*)

Fig. 2.9 Example of a
disconnected graph (**a**) and
hinged graph (**b**)

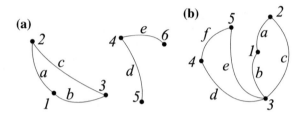

An example of a disconnected graph is shown in Fig. 2.9a.

A connected graph is **hinged** when it can be partitioned into two subgraphs connected by one node, called a *hinge*.

An example of a hinged graph is shown in Fig. 2.9b, where the hinge is node 3.

Loop: A subgraph containing only nodes of order 2, or a degenerate path where the first and last nodes are also of order 2, connected by an edge.

Figure 2.10 shows some examples of loops for the reference graph of Fig. 2.1b.

Mesh: A loop of a planar graph not containing any graph elements either inside (**inner loop**) or outside (**outer loop**).

Figure 2.11 shows some examples of meshes for the reference graph of Fig. 2.1b.

Cut-set: A set of edges of a graph which, when removed, make the graph disconnected.

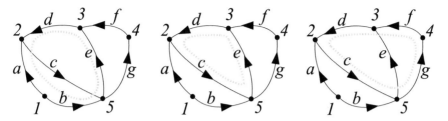

Fig. 2.10   Examples of loops (in *grey*)

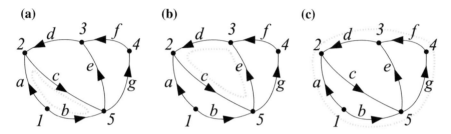

Fig. 2.11   Examples of meshes (in *grey*): inner loops (**a** and **b**) and outer loop (**c**)

As stated in Sect. 1.5.2, a cut-set can be easily associated with a closed path (or surface, for nonplanar graphs) crossing the cut-set edges. Actually, for each cut-set there are two possible closed paths, as shown in the examples of Fig. 2.12 for the reference graph of Fig. 2.1b.

> **Nodal cut-set**: A cut-set such that one of the two disconnected parts of the resulting graph is a single node.

Figure 2.13 shows some examples of nodal cut-sets for the reference graph of Fig. 2.1b.

2.1.3   Tree and Cotree

We now define the two basic graph structures used to find matrix formulations of Kirchhoff's laws.

> **Tree**: A subgraph containing all the N nodes and $N - 1$ edges of a given graph and in which any two nodes are connected by exactly one path.

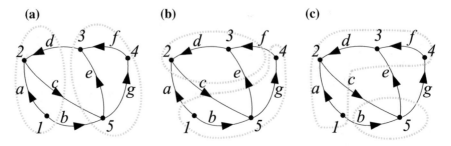

Fig. 2.12 Examples of cut-sets (corresponding to the *grey dashed* closed paths)

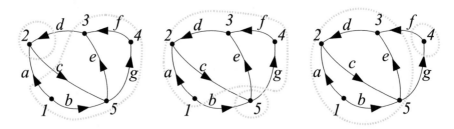

Fig. 2.13 Examples of nodal cut-sets (corresponding to the *grey dashed* closed paths)

Owing to this definition, a tree cannot contain any loop.

Cotree: A subgraph associated with a tree, containing all the N nodes and the $L - N + 1$ edges of the graph not contained in the tree.

Figure 2.14 shows some examples of trees and cotrees for the reference graph of Fig. 2.1b.

2.2 Matrix Formulation of Kirchhoff's Laws

As stated at the beginning of Sect. 2.1, these basic elements of graph theory can be used to formulate in a compact way (i.e., in matrix form) a set of independent Kirchhoff's laws for a given circuit. The goal is to find a complete[2] set of independent KVLs and KCLs, which are related to corresponding sets of independent loops and cut-sets, respectively. A set of independent loops (cut-sets) is also called a *basis of fundamental loops (cut-sets)*.

---

[2]The set is complete if any further KVL or KCL equation is linearly dependent on the equations belonging to the set.

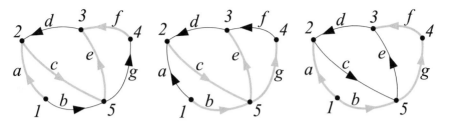

Fig. 2.14 Examples of trees (*thick grey edges*) and corresponding cotrees (*thin black edges*) for the reference graph of Fig. 2.1b

Fig. 2.15 Tree (*thick grey edges*) and cotree (*black edges*) for the reference graph of Fig. 2.1b

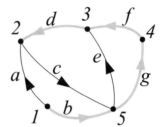

For planar graphs, the simplest choice for these bases is the set of $L - N + 1$ arbitrarily chosen meshes (which are independent loops) and the set of $N - 1$ arbitrarily chosen nodal cut-sets (which are independent cut-sets).

For generic graphs, a criterion to identify these bases refers to a tree and the corresponding cotree. In the following, we use the graph, tree, and cotree shown in Fig. 2.15. Moreover, henceforth I_q denotes the identity matrix of size q (i.e., the $q \times q$ square matrix with ones on the main diagonal and zeros elsewhere) and 0_q denotes the null column vector with q elements.

2.2.1 Fundamental Cut-Set Matrix

Each cut-set containing one and only one edge of the chosen tree is part of a basis of $(N - 1)$ fundamental cut-sets. Each fundamental cut-set is oriented (inwards/outwards) like the corresponding tree edge and is labeled as $\mathscr{C}_k$, where k denotes the edge. Figure 2.16 shows the basis of fundamental cut-sets for the considered example and the chosen tree.

Now, we can construct a matrix (of size $(N - 1) \times L$), called the *fundamental cut-set matrix*, where:

- Each row corresponds to exactly one fundamental cut-set (i.e., to the related tree edge).
- Each column corresponds to one graph edge. The columns are ordered as follows: first the cotree edges (ordered arbitrarily) and then the tree edges, in the same order

Fig. 2.16 Basis of
fundamental cut-sets for the
considered example

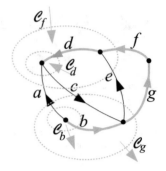

as for the rows. In the example, we follow the alphabetical order for both edge
sets.
- Each matrix entry is set to:

  0 If the edge on the column does not belong to the fundamental cut-set on the row
  1 If the edge on the column belongs to the fundamental cut-set on the row and has
  the same orientation
 −1 If the edge on the column belongs to the fundamental cut-set on the row and has
  the opposite orientation

In the considered example, the fundamental cut-set matrix is as follows.

$$
A = \begin{array}{c} \\ b \\ d \\ f \\ g \end{array}
\begin{array}{ccccccc} a & c & e & b & d & f & g \end{array}
\left(\begin{array}{ccc|cccc}
1 & 0 & 0 & 1 & 0 & 0 & 0 \\
1 & -1 & 0 & 0 & 1 & 0 & 0 \\
1 & -1 & 1 & 0 & 0 & 1 & 0 \\
1 & -1 & 1 & 0 & 0 & 0 & 1
\end{array} \right) = \left(\alpha | I_{N-1} \right) \qquad (2.1)
$$

We call i the column vector of descriptive currents associated with the oriented
edges of the graph and ordered exactly as are the columns of the cut-set matrix A;
that is, $i = (i_a \ i_c \ i_e \ i_b \ i_d \ i_f \ i_g)^T$. It is easy to check that the rows of A are linearly
independent; that is, the rank of A is $N - 1$. This is a general property, due to the
way the fundamental cut-set matrix is set up and to the fact that each row is related
to one element of a basis of cut-sets.

Of course, the cut-set orientation depends on the choice of the corresponding
closed path (as stated in Sect. 2.1.2), but the resulting matrix is invariant, as can be
easily checked.

Property

The system of equations

$$
Ai = 0_{N-1} \qquad (2.2)
$$

is **a set of** $N-1$ **independent KCLs for the circuit associated with the graph,** corresponding to the fundamental cut-sets related to the chosen tree.

For the circuit of Fig. 1.8 and for the choice of tree of Fig. 2.15, the set of independent KCLs is:

$$Ai = \begin{pmatrix} 1 & 0 & 0 & 1 & 0 & 0 & 0 \\ 1 & -1 & 0 & 0 & 1 & 0 & 0 \\ 1 & -1 & 1 & 0 & 0 & 1 & 0 \\ 1 & -1 & 1 & 0 & 0 & 0 & 1 \end{pmatrix} \begin{pmatrix} i_a \\ i_c \\ i_e \\ i_b \\ i_d \\ i_f \\ i_g \end{pmatrix} = \begin{pmatrix} i_a + i_b \\ i_a - i_c + i_d \\ i_a - i_c + i_e + i_f \\ i_a - i_c + i_e + i_g \end{pmatrix} = \begin{pmatrix} 0 \\ 0 \\ 0 \\ 0 \end{pmatrix} \quad (2.3)$$

Each row of the submatrix α contains information about the composition of the cut-set which the row refers to: for example, the nonzero elements in the row d of α indicate that $\mathscr{C}_d$ contains (in addition to d) the edges a and c; similarly, the row f of α indicates that $\mathscr{C}_f$ contains, in addition to f, the edges a, c, e.

We observe in passing that something similar can be observed for the columns of α: for example, the nonzero elements of the column a indicate that b, d, f, g are the tree edges forming a loop with a; similarly, the nonzero elements of the column c indicate that the tree edges d, f, g form a loop with c. Therefore α also contains topological information about the loops. This fact has major consequences on the fundamental loop matrix structure, discussed soon.

2.2.1.1 A Particular Case

For the specific tree choice shown in Fig. 2.17, we obtain the basis composed by nodal cut-sets only. Notice that the tree in this case is a *star subgraph*.

For this choice of tree, writing the cut-set matrix A according to the general rules, the set of independent KCLs is as follows.

$$Ai = \begin{pmatrix} 1 & 0 & 0 & 1 & 0 & 0 & 0 \\ -1 & -1 & 0 & 0 & 1 & 0 & 0 \\ 0 & -1 & 1 & 0 & 0 & 1 & 0 \\ 0 & 0 & -1 & 0 & 0 & 0 & 1 \end{pmatrix} \begin{pmatrix} i_a \\ i_d \\ i_f \\ i_b \\ i_c \\ i_e \\ i_g \end{pmatrix} = \begin{pmatrix} i_a + i_b \\ -i_a - i_d + i_c \\ -i_d + i_f + i_e \\ -i_f + i_g \end{pmatrix} = \begin{pmatrix} 0 \\ 0 \\ 0 \\ 0 \end{pmatrix} \quad (2.4)$$

Fig. 2.17 Choice of tree
(*thick grey edges*)
corresponding to a basis of
nodal cut-sets only (*dashed
lines*)

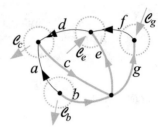

Fig. 2.18 Basis of
fundamental loops for the
considered example

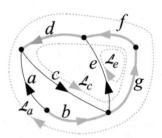

This set of equations is completely equivalent to Eq. 2.3.

This matrix is strictly related to the so-called *incidence matrix*.

You can check your comprehension by obtaining Eq. 2.4 through linear combinations of Eq. 2.3.

2.2.2 Fundamental Loop Matrix

Each loop containing only one edge of a cotree is part of a basis of $(L - N + 1)$ fundamental loops. Each fundamental loop is oriented as is the corresponding cotree edge and is labeled as $\mathcal{L}_k$, where k denotes the cotree edge. Figure 2.18 shows the basis of fundamental loops for the considered example and the chosen tree.

Now, we can construct a matrix (of size $(L - N + 1) \times L$), called the *fundamental loop matrix*, where:

- Each row corresponds to exactly one fundamental loop (i.e., to the related cotree edge).
- Each column corresponds to one graph edge. The columns are ordered as in matrix *A*.
- Each matrix entry is set to:

  0 If the edge on the column does not belong to the fundamental loop on the row
  1 If the edge on the column belongs to the fundamental loop on the row and has the same orientation
  −1 If the edge on the column belongs to the fundamental loop on the row and has the opposite orientation

In the considered example, the fundamental loop matrix is:

$$B = \begin{array}{c} \\ a \\ c \\ e \end{array} \begin{array}{ccc|cccc} a & c & e & b & d & f & g \\ \left(1 & 0 & 0 \right. & -1 & -1 & -1 & -1 \\ 0 & 1 & 0 & 0 & 1 & 1 & 1 \\ 0 & 0 & 1 & 0 & 0 & -1 & -1 \end{array} \left. \right) = \left(I_{L-N+1} | -\alpha^T \right) \qquad (2.5)$$

We call v the column vector of descriptive voltages associated with the oriented edges of the graph and ordered exactly as are the columns of the loop matrix B; that is, $v = (v_a \ v_c \ v_e \ v_b \ v_d \ v_f \ v_g)^T$. It is easy to check that the rows of B are linearly independent; this is a general property, due to the way the fundamental loop matrix is set up and to the fact that each row is related to one element of a basis of loops. For this reason, the rank of B is $L - N + 1$. When, as in this case, the ordering of the tree edges is the same for the matrices A and B, the elements of A and B are related very simply: the matrix part complementary to the identity submatrix is α in A and $-\alpha^T$ in B. This follows from the previously observed property concerning the columns of α; that is, for any column j of α, the tree edges i with $\alpha_{ij} \neq 0$ are the constituents of the loop $\mathscr{L}_j$.

Property

The system of equations

$$Bv = 0_{L-N+1} \qquad (2.6)$$

is **a set of $L - N + 1$ independent KVLs for the circuit associated with the graph**, corresponding to the fundamental loops related to the chosen tree.

For the circuit of Fig. 1.8 and for the choice of tree of Fig. 2.15, the set of independent KVLs is as follows.

$$Bv = \begin{pmatrix} 1 & 0 & 0 & -1 & -1 & -1 & -1 \\ 0 & 1 & 0 & 0 & 1 & 1 & 1 \\ 0 & 0 & 1 & 0 & 0 & -1 & -1 \end{pmatrix} \begin{pmatrix} v_a \\ v_c \\ v_e \\ v_b \\ v_d \\ v_f \\ v_g \end{pmatrix} = \begin{pmatrix} v_a - v_b - v_d - v_f - v_g \\ v_c + v_d + v_f + v_g \\ v_e - v_f - v_g \end{pmatrix} = \begin{pmatrix} 0 \\ 0 \\ 0 \end{pmatrix}$$

$$(2.7)$$

2.2.2.1 A Particular Case

For the star tree shown in Fig. 2.19, we obtain the basis composed by all the inner loops.

Fig. 2.19 Choice of tree
(*thick grey edges*)
corresponding to the basis of
all the inner loops (*dashed
loops*)

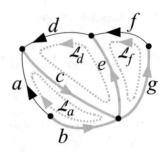

For this choice of tree, the set of independent KVLs is:

$$Bv = \begin{pmatrix} 1 & 0 & 0 & -1 & 1 & 0 & 0 \\ 0 & 1 & 0 & 0 & 1 & 1 & 0 \\ 0 & 0 & 1 & 0 & 0 & -1 & 1 \end{pmatrix} \begin{pmatrix} v_a \\ v_d \\ v_f \\ v_b \\ v_c \\ v_e \\ v_g \end{pmatrix} = \begin{pmatrix} v_a - v_b + v_c \\ v_d + v_c + v_e \\ v_f - v_e + v_g \end{pmatrix} = \begin{pmatrix} 0 \\ 0 \\ 0 \end{pmatrix} \qquad (2.8)$$

This set of equations is completely equivalent to Eq. 2.7. You can check your comprehension by obtaining Eq. 2.8 through linear combinations of Eq. 2.7.

2.2.3   Some General Concepts on Vector Spaces and Matrices

A *vector space* $\mathcal{V}$ is a nonempty set of vectors such that, for any two vectors x_1 and x_2 of $\mathcal{V}$, any of their linear combinations $\beta_1 x_1 + \beta_2 x_2$ ($\beta_1, \beta_2 \in \mathbb{R}$) is still an element of $\mathcal{V}$. The null element 0 is always a vector of $\mathcal{V}$.

The *dimension* of $\mathcal{V}$, denoted as dim($\mathcal{V}$), is the maximum number of linearly independent vectors in $\mathcal{V}$ and must not be confused with the number of components of the elements of $\mathcal{V}$.

A set of linearly independent vectors in $\mathcal{V}$ consisting of dim($\mathcal{V}$) vectors is called a *basis* for $\mathcal{V}$.

Given p vectors $x_1, \ldots, x_p$ with the same number of components, the set of all linear combinations $\sum_{i=1}^{p} \beta_i x_i$ is a vector space called the *span* of these vectors. For instance, the vector space $\mathcal{V}$ is the span of dim($\mathcal{V}$) linearly independent vectors. The span of a number of linearly independent vectors lower than dim($\mathcal{V}$) generates a *subspace* $\mathcal{L}$ of $\mathcal{V}$. For instance, Fig. 2.20 shows an example for $\mathcal{V} \equiv \mathbb{R}^3$.

The vectors x_1 and x_2 (as well as all their linear combinations $\beta_1 x_1 + \beta_2 x_2$, with $\beta_1, \beta_2 \in \mathbb{R}$) lie in a plane $\mathcal{L}$, which is a two-dimensional subspace of $\mathbb{R}^3$ passing through the origin.

Fig. 2.20 A
two-dimensional subspace
$\mathscr{L}$ in $\mathbb{R}^3$

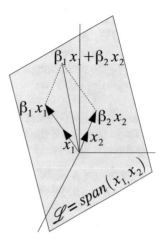

As stated above, it is important not to confuse the dimension of the vector space (or subspace) with the number of components (the size) of its individual vectors, because they are not necessarily the same. In the considered example, for instance, the vectors x_1 and x_2 have three components, despite their belonging to the two-dimensional subspace $\mathscr{L}$.

In the following, we introduce some specific spaces and subspaces associated with a matrix [3, 4], in order to provide (in the next section) a geometrical interpretation of the matrix formulation of Kirchhoff's laws, thus settling the basis for introducing Tellegen's theorem.

Let us consider a matrix $Q \in \mathbb{R}^{m \times n}$. We can write Q in terms of its columns as $Q = (q_1 \ldots q_n)$. Let x denote any vector in $\mathbb{R}^n$. The vector space

$$\mathscr{R}(Q) = \left\{ y \in \mathbb{R}^m : y = Qx, \ x \in \mathbb{R}^n \right\}$$

is called the *range* of Q. We can also write, in terms of the column vectors q_i,

$$\mathscr{R}(Q) = span \, (q_1, \ldots, q_n) \, .$$

In the general case, the linearly independent columns of Q can be a subset of $\{q_1, q_2, \ldots, q_n\}$. It can be shown that the maximum number of linearly independent columns of Q and the maximum number of its linearly independent rows are equal. This common value r is the *rank* of Q. Then $rank(Q) = rank(Q^T) = r \leq \min(m, n)$ and $\dim(\mathscr{R}(Q)) = r$.

The set of all solutions to the homogeneous system $Qz = 0$,

$$\mathscr{N}(Q) = \left\{ z \in \mathbb{R}^n : Qz = 0 \right\}$$

is called the *null space* of Q (or *kernel* of Q).

In the same way we can define the vector spaces associated with the transpose of Q: $\mathscr{R}(Q^T)$, $\mathscr{N}(Q^T)$.

Two m-size vectors $w_R \in \mathscr{R}(Q)$ and $w_0 \in \mathscr{N}(Q^T)$ are always orthogonal; owing to the definition of $\mathscr{R}(Q)$, there must exist a vector $\bar{x}$ such that $w_R = Q\bar{x}$, thus $w_R^T w_0 = (Q\bar{x})^T w_0 = \bar{x}^T \underbrace{Q^T w_0}_{0} = 0$. An analogous result holds for two n-size vectors $x_R \in \mathscr{R}(Q^T)$ and $x_0 \in \mathscr{N}(Q)$.

These spaces are the main ingredients of two important results concerning the decomposition of vectors:

1 Any vector $x \in \mathbb{R}^n$, the domain space of Q, can be uniquely decomposed as $x = x_R + x_0$, where $x_0 \in \mathscr{N}(Q)$ and $x_R \in \mathscr{R}(Q^T)$. Then $\mathscr{N}(Q)$ and $\mathscr{R}(Q^T)$ are complementary and disjoint ($\mathscr{N}(Q) \cap \mathscr{R}(Q^T) = \emptyset$, empty set) subspaces of $\mathbb{R}^n$; that is, $\mathbb{R}^n$ is given by the *direct sum* ($\oplus$) of the two subspaces:

$$\mathbb{R}^n = \mathscr{N}(Q) \oplus \mathscr{R}(Q^T) \text{ and } n = \dim(\mathscr{N}(Q)) + r.$$

The subspace $\mathscr{N}(Q)$ is an empty set if and only if $r = n$.

2 Any vector $w \in \mathbb{R}^m$, the codomain space of Q, can be uniquely decomposed as $w = w_R + w_0$, where $w_0 \in \mathscr{N}(Q^T)$ and $w_R \in \mathscr{R}(Q)$. Then $\mathscr{N}(Q^T)$ and $\mathscr{R}(Q)$ are complementary and disjoint ($\mathscr{N}(Q^T) \cap \mathscr{R}(Q) = \emptyset$) subspaces of $\mathbb{R}^m$; that is, $\mathbb{R}^m$ is given by the direct sum of the two subspaces

$$\mathbb{R}^m = \mathscr{N}(Q^T) \oplus \mathscr{R}(Q) \text{ and } m = \dim(\mathscr{N}(Q^T)) + r.$$

The subspace $\mathscr{N}(Q^T)$ is an empty set if and only if $r = m$.

To exemplify the above concepts, let us consider the matrix

$$Q = \begin{matrix} & q_1 & q_2 \\ & \begin{pmatrix} 2 & 0 \\ 0 & 1 \\ 1 & 1 \end{pmatrix} \end{matrix}$$

which has $m = 3$, $n = 2$ and rank $r = 2$. Its column vectors q_1, q_2 define the plane $\mathscr{R}(Q)$:

$$\mathscr{R}(Q) = span(q_1, q_2) = Q \begin{pmatrix} \beta_1 \\ \beta_2 \end{pmatrix} = \beta_1 q_1 + \beta_2 q_2; \qquad \beta_1, \beta_2 \in \mathbb{R}.$$

Taking as reference the orthogonal directions a_1, a_2, a_3, the vectors q_1, q_2 are shown in Fig. 2.21. The plane $\mathscr{R}(Q)$ intersects the $a_1 a_3$-plane along the line of q_1 and the $a_2 a_3$-plane along the line of q_2.

Fig. 2.21 The plane $\mathscr{R}(Q)$ and the line $\mathscr{N}(Q^T)$ associated with the (3×2) matrix Q

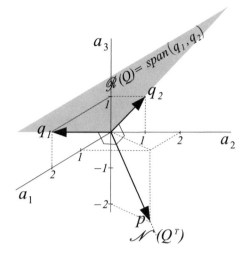

Because $\dim(\mathscr{N}(Q^T)) = m - r = 1$, the complementary subspace $\mathscr{N}(Q^T)$ is a straight line orthogonal to the plane $\mathscr{R}(Q)$. Denoting as $p = (p_1 \; p_2 \; p_3)^T$ a vector along this line, we have

$$Q^T p = \begin{pmatrix} 2 & 0 & 1 \\ 0 & 1 & 1 \end{pmatrix} \begin{pmatrix} p_1 \\ p_2 \\ p_3 \end{pmatrix} = \begin{pmatrix} 0 \\ 0 \end{pmatrix} \quad \Rightarrow \quad \begin{cases} 2p_1 + p_3 = 0 \\ p_2 + p_3 = 0 \end{cases}$$

Then the components p_1, p_2 can be expressed in terms of p_3, that parameterizes the points of the subspace. The vector p plotted in the figure corresponds to $p_3 = -2$.

Finally, inasmuch as $r = n$, we have $\dim(\mathscr{N}(Q)) = 0$ (empty subspace) and $\mathscr{R}(Q^T) = \mathbb{R}^n$.

2.2.4 The Cut-Set and Loop Matrices and Their Associate Space Vectors

Consider a directed graph with L edges and N nodes. This graph can be arbitrarily partitioned into a tree and its cotree. Such a partition leads to the definition of a cut-set matrix A and a loop matrix B, as shown in Sects. 2.2.1 and 2.2.2. A current vector i and a voltage vector v, both of size L, are said to be *compatible* with the graph if they

Fig. 2.22 Matrix
A-relationships between
spaces for compatible
voltage and current vectors

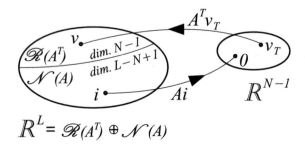

$$R^L = \mathscr{R}(A^T) \oplus \mathscr{N}(A)$$

satisfy the KCLs and KVLs, respectively, that is, if $Ai = 0$ and $Bv = 0$. Because
the structure of A (where $m = N - 1$ and $n = L$) is $(\alpha | I_{N-1})$, from $Ai = 0$ we have
$N - 1$ independent scalar equations, which represent as many constraints on the L
elements of the vector i. Therefore, due to KCLs, the number of degrees of freedom
for the current elements of a vector i compatible with the graph is $L - N + 1$.

In terms of vector spaces, $Ai = 0$ means that i belongs to $\mathscr{N}(A)$, the null space
of matrix A, whose dimension is $L - N + 1$.

Consider now the KVLs $Bv = 0$, with $B = (I_{L-N+1} | -\alpha^T)$ (where $m = L-N+1$
and $n = L$). The vector v can be partitioned into two subvectors v_C and v_T, which
contain the $L - N + 1$ voltages on the cotree edges and the $N - 1$ voltages on the
tree edges, respectively:

$$v = \begin{pmatrix} v_C \\ v_T \end{pmatrix} \tag{2.9}$$

Owing to this partition, the KVLs $Bv = 0$ can be recast as $I_{L-N+1}v_C - \alpha^T v_T = 0$;
that is, $v_C = \alpha^T v_T$. Then, we directly obtain:

$$v = \begin{pmatrix} v_C \\ v_T \end{pmatrix} = \begin{pmatrix} \alpha^T \\ I_{N-1} \end{pmatrix} v_T = A^T v_T. \tag{2.10}$$

It follows that each vector v of voltages compatible with the graph can be obtained
through a product $A^T v_T$. This means that $v \in \mathscr{R}(A^T)$, whose dimension is $N-1$. The
values of the $N - 1$ components of the subvector v_T can be assigned independently,
therefore the voltage elements of a compatible vector v can be chosen with $N - 1$
degrees of freedom, due to KVLs, which impose $L - N + 1$ constraints on the L
components of v.

Figure 2.22 summarizes all these results and also highlights the roles of the matrices A and A^T as operators for the passage between the subspaces of $\mathbb{R}^L$ and the
space $\mathbb{R}^{N-1}$.

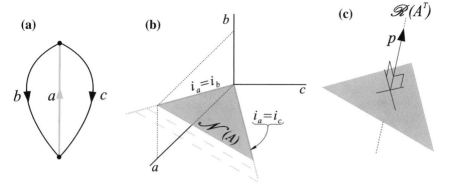

Fig. 2.23 Case Study: **a** graph; **b, c** spaces $\mathcal{N}(A)$ and $\mathcal{R}(A^T)$ for compatible current and voltage vectors

Case Study

Consider the very simple graph ($L = 3$, $N = 2$) shown in Fig. 2.23a. Taking the edge a as the (only) tree edge and the edges b, c as cotree edges, the fundamental cut-set matrix A is

$$\begin{array}{ccc} & b & c & a \end{array}$$
$$A = a \ \begin{pmatrix} -1 & -1 & | & 1 \end{pmatrix} \tag{2.11}$$

Therefore, KCL reduces to a single scalar equation:

$$i = \begin{pmatrix} i_b \\ i_c \\ i_a \end{pmatrix}; \quad Ai = -i_b - i_c + i_a = 0 \tag{2.12}$$

The three components i_b, i_c, i_a of any current vector i compatible with the graph must fulfill the KCL constraint $i_a = i_b + i_c$, which leads to the expression for the two-dimensional subspace $\mathcal{N}(A)$:

$$i = \begin{pmatrix} i_b \\ i_c \\ i_b + i_c \end{pmatrix} = i_b \begin{pmatrix} 1 \\ 0 \\ 1 \end{pmatrix} + i_c \begin{pmatrix} 0 \\ 1 \\ 1 \end{pmatrix}. \tag{2.13}$$

In the above expression, the values of i_b, i_c play the role of span coefficients.

Denoting by b, c, a the orthogonal directions spanning the $\mathbb{R}^3$ space as shown in Fig. 2.23b, $\mathcal{N}(A)$ is the plane that intersects the plane $i_c = 0$ along the straight line $i_a = i_b$ and the plane $i_b = 0$ along the straight line $i_a = i_c$. All the vectors $i \in \mathbb{R}^3$ compatible with the graph lie on the plane $\mathcal{N}(A)$.

The voltage vector

$$v = \begin{pmatrix} v_b \\ v_c \\ v_a \end{pmatrix} \tag{2.14}$$

can be partitioned according to Eq. 2.9; in particular, we have $v_T = v_a$. With this in mind, and recalling Eq. 2.10, any vector v compatible with the graph can be obtained as

$$v = A^T v_T = \begin{pmatrix} -1 \\ -1 \\ 1 \end{pmatrix} v_a \tag{2.15}$$

or, in a more general formulation highlighting the parametric role of the term v_a, as

$$v = p\beta; \quad p = \begin{pmatrix} -1 \\ -1 \\ 1 \end{pmatrix}; \quad \beta \in \mathbb{R}. \tag{2.16}$$

Therefore, any vector v such that $Bv = 0$ is proportional to the vector p. It is easy to verify that p is orthogonal to any vector $i \in \mathcal{N}(A)$, as shown in Fig. 2.23c. The way to prove it is based on the observation that, being $i_a = i_b + i_c$, we can write i as $(i_b \ i_c \ (i_b + i_c))^T$ and then:

$$p^T i = (-1 \ -1 \ 1) \begin{pmatrix} i_b \\ i_c \\ i_b + i_c \end{pmatrix} = 0. \tag{2.17}$$

You can check the correspondence of these results with the general ones shown in Fig. 2.22.

In a similar fashion, denoting by i_C and i_T the subvectors containing, respectively, the $L - N + 1$ cotree currents and the $N - 1$ currents through the tree edges, we have

$$i = \begin{pmatrix} i_C \\ i_T \end{pmatrix} \tag{2.18}$$

which enables us to recast the KCLs $Ai = 0$ as $\alpha i_C + I_{N-1} i_T = 0$; that is, $i_T = -\alpha i_C$. With this in mind, we obtain

$$i = \begin{pmatrix} i_C \\ i_T \end{pmatrix} = \begin{pmatrix} I_{L-N+1} \\ -\alpha \end{pmatrix} i_C = B^T i_C. \tag{2.19}$$

Therefore, each current vector i compatible with the graph can be obtained through a product $B^T i_C$. This means that the current elements of any compatible vector i can be chosen with the $L - N + 1$ degrees of freedom representing the size of the subvector

Fig. 2.24 Matrix
B-relationships between
spaces for compatible
voltage and current vectors

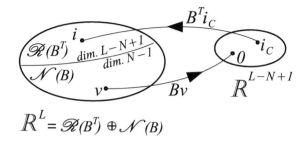

$$R^L = \mathscr{R}(B^T) \oplus \mathscr{N}(B)$$

i_C; moreover, $i \in \mathscr{R}(B^T)$. Because $Bv = 0$ means that v belongs to $\mathscr{N}(B)$, the space R^L can be thought of as partitioned into the two subspaces $\mathscr{R}(B^T)$ and $\mathscr{N}(B)$. This partition is shown in Fig. 2.24, which highlights the roles of the matrices B and B^T as operators for the passage between the subspaces of R^L and the space R^{L-N+1}.

The properties presented in this section are the basis for Tellegen's theorem, which is treated in the next section.

2.3 Tellegen's Theorem

Theorem 2.1 (Tellegen's theorem) *In a directed graph, any compatible voltage vector v is orthogonal to any compatible current vector i.*

Proof To prove this, just consider that, thanks to the compatibility assumption, we have

$$v^T i = (A^T v_T)^T i = v_T^T \underbrace{Ai}_{0} = 0. \tag{2.20}$$

$\square$

Tellegen's theorem is one of the most general theorems of circuit theory [5]. It depends only on Kirchhoff's laws and on the circuit's topology (graph), and it holds regardless of the physical nature of the circuit's components or the waveforms of voltages and currents, and so on. Therefore the voltages and currents that are used for Tellegen's theorem are not necessarily those actually present in a given circuit. By introducing specific assumptions about the physical properties of the components, waveforms and so on, Tellegen's theorem is the starting point to obtain, usually in a direct way, various specific and useful results. In the next chapters we show that for many circuit properties, the proof that can be given by relying on Tellegen's theorem is simpler than others and its range of validity is more clearly demonstrated.

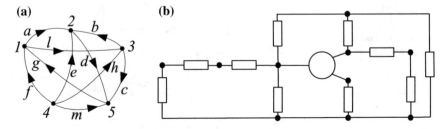

Fig. 2.25 Problems 2.1 (**a**) and 2.2 (**b**)

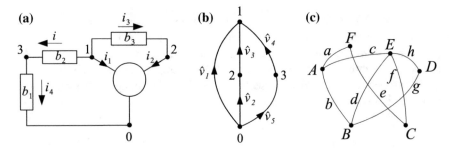

Fig. 2.26 Problems 2.3 (**a, b**) and 2.4 (**c**)

2.4   Problems

2.1 Choose a tree for the nonplanar graph shown in Fig. 2.25a and find the corresponding fundamental cut-set and loop matrices.

2.2 Determine the number of KCLs and KVLs necessary to solve the circuit shown in Fig. 2.25b. **Hint**: Consider the component connections to the lowest wire as a single node (dot).

2.3 Assume that you can measure the voltages of another circuit whose graph is shown in Fig. 2.26b. Is it possible to determine current i in Fig. 2.26a by measuring current i_3 in the same circuit? How?

2.4 Determine the number of fundamental loops, fundamental cut-sets, and tree edges for the graph shown in Fig. 2.26c.

References

1. Euler L (1741) Solutio Problematis ad Geometriam Situs Pertinentis. Commentarii Academiae Scientiarum Imperialis Petropolitanae 8:128–140
2. Kuratowski C (1930) Sur le problème des courbes gauches en topologie. Fund. Math. 15:271–283
3. Meyer CD (2000) Matrix analysis and applied linear algebra. SIAM, Philadelphia
4. Laub AJ (2005) Matrix analysis for scientists and engineers. SIAM, Philadelphia
5. Penfield P, Spence R, Duinker S (1970) Tellegen's theorem and electrical networks. MIT Press, Cambridge

Part II
Memoryless Multi-terminals: Descriptive Equations and Properties

Chapter 3
Basic Concepts

.

The important thing is not to stop questioning. Curiosity has its own reason for existing.

A. Einstein

Abstract In Chap. 1, we introduced circuit equations (KCL and KVL) related only to the way the circuit components are connected (the circuit *topology*). These equations are completely independent of the components' nature, that is, of their physical behavior. In this chapter we begin considering the specific physical role played by each component, which is described through the so-called *descriptive equation(s)*. Then, we introduce the descriptive equations of electrical and electronic two-terminal components (resistor, voltage and current sources, diode) and define general component characteristics and properties: implicit and explicit representations, linearity, time-invariance, memory, energy-based classification, and reciprocity. Some introductory examples of circuit analysis are proposed. Then the Thévenin and Norton equivalent representations of two-terminal resistive components are introduced. The main connections of two-terminal resistive elements conclude the chapter.

3.1 Solving a Circuit: Descriptive Versus Topological Equations

Consider a circuit made up of M components. Denoting by n_k the number of terminals of the kth component ($k = 1, \ldots, M$), the circuit has

$$L = \sum_{k=1}^{M}(n_k - 1) \tag{3.1}$$

descriptive voltages and L descriptive currents.

© Springer International Publishing AG 2018

M. Parodi and M. Storace, *Linear and Nonlinear Circuits:*
Basic & Advanced Concepts, Lecture Notes in Electrical Engineering 441,
DOI 10.1007/978-3-319-61234-8_3

The determination of these $2L$ descriptive variables requires $2L$ equations, which are of two types: *descriptive equations* and *topological equations*. Solving a circuit means using at most $2L$ equations to find the value (or the symbolic expression) of one (or more) of these variables.

The descriptive equations derive from the physical behavior of each component, that is, from its specific nature. An n_k-terminal component has $(n_k - 1)$ descriptive equations that, in general, involve all the $2(n_k - 1)$ descriptive variables of the component itself. For example, any two-terminal component has a single descriptive equation; a three-terminal component needs two descriptive equations, and so on. Denoting by v the vector of descriptive voltages: $v = \left(v_1 \cdots v_{n_k-1}\right)^T$ and by i the analogous vector of descriptive currents: $i = \left(i_1 \cdots i_{n_k-1}\right)^T$, the descriptive equations can be written in *implicit form* as

$$f_j(v, i) = 0 \quad (j = 1, \ldots, n_k - 1). \tag{3.2}$$

The term f_j represents a (scalar) algebraic function of its arguments, but in other cases it may contain derivatives or integrals with respect to time. In this chapter, we introduce only components described by algebraic f_j functions.

Because the number of descriptive equations for an n_k-terminal is $(n_k - 1)$, the total number of these equations in the circuit is L. Therefore, the number of necessary topological equations also amounts to L. If the circuit has N nodes and L edges, it has been shown in Sects. 2.2.2 and 2.2.1 that the KVLs provide $(L - N + 1)$ independent equations, whereas the KCLs provide the remaining $(N - 1)$ ones. Summing up, the $2L$ equations necessary to solve the circuit are generated according to the diagram of Fig. 3.1: *the necessary and sufficient information for the complete study of any circuit, for half originate from its topological structure (the way in which its components are connected) and for half by the components' physical characteristics.*

Fig. 3.1 The two types of circuit equations

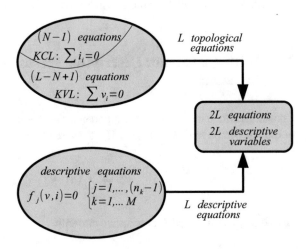

3.2 Descriptive Equations of Some Components

Here we introduce the descriptive equations for some largely used components.

3.2.1 Resistor

The resistor is an electrical device (some examples are shown in Fig. 3.2a), whose two-terminal model is shown in Fig. 3.2b.

> The **resistor descriptive equation** – also called *Ohm's law* –[1] is
>
> $$v(t) = Ri(t) \qquad\qquad (3.3)$$

R is a parameter called *(electrical) resistance*. The (derived) SI unit of measurement of resistance is the *ohm*, whose symbol is Ω. According to Ohm's law, it is evident that $[\Omega] = [V/A]$.

The implicit form of Ohm's law is $f(i, v) = v - Ri = 0$. In a physical resistor, the resistance R is positive and its value depends on many factors, mainly temperature and age. The resistor model, instead, assumes that R is constant. Moreover, its value is usually positive, but in some cases it can also be negative, to model more complex physical devices, as we show later.

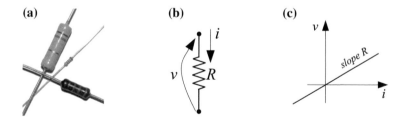

(a)   **(b)**   **(c)**

Fig. 3.2 Resistor: **a** three physical devices (the band colors code the resistance values); **b** model; **c** graphical representation of Ohm's law

---

[1]The law was named after the German physicist Georg Simon Ohm (1789–1854), who, in a treatise published in 1827, presented a slightly more complex equation than the one above to explain his experimental results about measurements of applied voltage and current through simple electrical circuits containing wires of various lengths. The above equation is the modern form of Ohm's law.

Fig. 3.3 Limit cases: **a** short
circuit ($R = 0$); **b** open
circuit ($G = 0$)

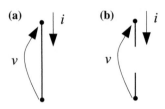

Fig. 3.4 Voltage source: **a**
model; **b** graphical
representation of the
descriptive equation

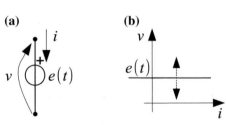

The inverse of the resistance $G = 1/R$ is called the *conductance*. Its (derived) SI
unit of measurement is the *siemens*, whose symbol is S.[2] Of course, $[S] = [\Omega^{-1}] =$
$[A/V]$.

The descriptive equation is linear and homogeneous, therefore it is represented on
the (i, v) plane by a straight line with slope R passing through the origin, as shown
in Fig. 3.2c.

3.2.1.1   Limit Cases

For $R = 0$, the descriptive equation becomes $v(t) = 0$. In this case, the two-terminal
is called a *short circuit* and is represented by the model shown in Fig. 3.3a. For
$G = 0$ (i.e., $R \rightarrow \infty$), the descriptive equation becomes $i(t) = 0$. In this case,
the two-terminal is called an *open circuit* and is represented by the model shown in
Fig. 3.3b.

3.2.2   *Ideal Voltage Source*

The (ideal) voltage source (shown in Fig. 3.4a) is a two-terminal that imposes a
prescribed voltage, say $e(t)$, between the two nodes.

---

[2]The unit is named after Ernst Werner von Siemens (1816–1892), a German inventor and industri-
alist.

The **voltage source equation** is

$$v(t) = e(t) \tag{3.4}$$

$e(t)$ is called the *impressed voltage* and is measured in volts.

The implicit form of the descriptive equation is $f(i, v) = v - e = 0$. We remark that current i does not appear in the equation: this means that its value depends on the rest of the circuit (and not that its value is 0A, as sometimes supposed by imaginative students).

The most common expressions for the impressed voltage are two:

- $e(t) = E$: In this case, the *direct current (DC) source*, the voltage is constant and the voltage source models a battery. Of course, a real battery impresses a constant voltage only for a limited period of time.
- $e(t) = E \cos(\omega t + \phi) = E \cos(2\pi F t + \phi)$: In this case, *alternating current (AC) source*, the voltage oscillates in time and the voltage source models an electric socket. Parameter E is a voltage amplitude (measured in volts), ϕ is an angle (expressed in radians), and F is a frequency (measured in s^{-1} or in hertz[3]) called *utility frequency, (power) line frequency*, or *mains frequency*. Worldwide, typical parameter values are $F = 50\,$Hz (with amplitudes E ranging from 220 to 240 V) and $F = 60\,$Hz (with amplitudes E ranging from 100 to 127 V).

Inasmuch as the descriptive equation is in general not homogeneous and not constant, it is represented on the (i, v) plane by a horizontal straight line moving vertically with time, as shown in Fig. 3.4b.

In the limit case $e(t) = 0$, the voltage source is turned off and is equivalent to a short circuit.

3.2.3 Ideal Current Source

The (ideal) current source (shown in Fig. 3.5a) is a two-terminal that generates a prescribed current, say $a(t)$.

The **current source equation** is

$$i(t) = a(t) \tag{3.5}$$

$a(t)$ is called the *impressed current* and is measured in amperes.

---

[3]The (derived) SI unit of measurement of frequency is the hertz (whose symbol is Hz), so named in honor of Heinrich Rudolf Hertz (1857–1894), the German physicist who first conclusively proved the existence of electromagnetic waves theorized by James Clerk Maxwell (1831–1879).

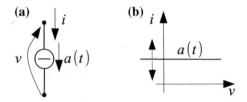

Fig. 3.5 Current source: **a** model; **b** graphical representation of the descriptive equation

The implicit form of the descriptive equation is $f(i, v) = i - a = 0$. We remark that voltage v does not appear in the equation: this means that its value depends on the rest of the circuit.

The most common expressions for the impressed current are two: $a(t) = A$ (DC source) and $a(t) = A \cos(\omega t + \phi) = A \cos(2\pi F t + \phi)$ (AC source).

Because the descriptive equation is in general not homogeneous and not constant, it is represented on the (v, i) plane by a horizontal straight line moving vertically with time, as shown in Fig. 3.5b.

In the limit case $a(t) = 0$, the current source is turned off and is equivalent to an open circuit.

3.2.4 Elementary Circuits

Here we provide five examples of the analysis of simple circuits. As a general remark (according to Ockham's razor), we always aim to solve the circuit by involving *the minimum number of unknowns*, in addition to those requested by the problem. This allows us to reduce the number of equations and unknowns to be handled, thus also decreasing the sources of errors. Working on the circuit scheme is a great help from this standpoint. As shown, in most circuits we find the final solution by handling few equations.

Remark Another great help to prevent errors is to check the physical dimensions in the obtained symbolic expressions.

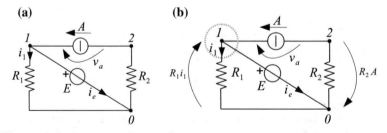

Fig. 3.6 Case Study 1

Case Study 1

Find the symbolic expressions and the numerical values for i_1, i_e, and v_a in the circuit shown in Fig. 3.6a, with $E = 2V$, $A = 1mA$, $R_1 = 5k\Omega$, $R_2 = 7k\Omega$.

Notice that for the current source the descriptive variables are not taken according to the standard choice. This is not a concern if we are not required to compute the power absorbed/delivered by this component.

We can start by finding the variables that can be expressed in terms of the given unknowns: the voltage on R_1 (taken according to the standard choice) is $R_1 i_1$ and the voltage on R_2 (taken according to the standard choice) is $R_2 A$. (See Fig. 3.6b.) Then, by using the voltage source equation and the KVL for the left mesh, we have $R_1 i_1 = E$; that is, $i_1 = E/R_1 = 2/(5 \cdot 10^3)$ A $= 0.4$ mA $= 400\mu$ A. By using the voltage source equation and the KVL for the right mesh, we have $v_a = E + R_2 A = [2 + (7 \cdot 10^3)(1 \cdot 10^{-3})]$V $= 9$ V. Finally, by using the KCL at nodal cut-set 1 (Fig. 3.6b), we obtain $A = i_1 + i_e$; that is, $i_e = A - i_1 = (1 \cdot 10^{-3} - 0.4 \cdot 10^{-3})$ A $= 0.6$mA $= 600\mu$ A.

As an alternative, the same circuit can be solved by introducing all possible descriptive variables ($2L = 8$, in this case), as shown in Fig. 3.7, and solving a system of eight equations. ($N - 1 = 2$ KCLs, $L - N + 1 = 2$ KVLs, and $L = 4$ descriptive equations.)

For instance, we can take the two inner loops in Fig. 3.7 to find the KVLs $v_1 = v_e$ and $v_a + v_2 = v_e$. Similarly, we can take the two nodal cut-sets 1 and 2 to find the KCLs $i_1 + i_e + i_a = 0$ and $i_a = i_2$. The four descriptive equations are: $v_1 = R_1 i_1$, $v_2 = R_2 i_2$, $v_e = E$, and $i_a = -A$. (Pay attention to the current directions!) By solving this system, one finds all the descriptive variables, included the requested ones.

The first solving method in practice shortens the process of substitution of the unknowns not requested by the problem. Even for a very simple circuit such as this, it is evident that managing a set of $2L$ unknowns can increase the probability of errors, therefore we recommend once more the introduction of the minimum number of unknowns, as in the first solution.

Fig. 3.7 All descriptive
variables for Case Study 1

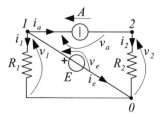

Fig. 3.8 Case Study 2

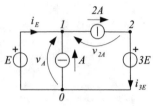

Case Study 2

Find the symbolic expressions and the numerical values for the currents flowing through the voltage sources and the voltages across the current sources in the circuit shown in Fig. 3.8, with $E = 2V$, $A = 3mA$. Compute the powers delivered by the current sources and absorbed by the voltage sources.

From the left mesh, we immediately have (KVL) $v_A = E = 2$ V. Then, the power delivered by the current source A is $p_A = EA = 2 \cdot 3 \ 10^{-3}$ W $= 6$ mW. From the outer loop, we have (KVL) $E = v_{2A} + 3E$; that is, $v_{2A} = -2E$. Thus the power delivered by the current source $2A$ is $p_{2A} = -v_{2A} \ 2A = 4EA = 4 \cdot 2 \cdot 3 \ 10^{-3}$ W $= 24$ mW. From KCL at node 2, $i_{3E} = 2A = 2 \cdot 3 \ 10^{-3}$A $= 6$ mA and the power absorbed by the voltage source $3E$ is $p_{3E} = 3Ei_{3E} = 6EA = 6 \cdot 2 \cdot 3 \ 10^{-3}$ W $= 36$ mW. From KCL at node 1, $i_E + A = 2A$; that is, $i_E = A = 3$ mA and the power absorbed by the voltage source E is $p_E = E(-i_E) = -EA = -2 \cdot 3 \ 10^{-3}$ W $= -6$ mW.

Remark: The total power absorbed by the circuit is $-p_A - p_{2A} + p_{3E} + p_E = 0$ W, according to Tellegen's theorem and to the law of conservation of energy. (See Sect. 2.3.)

Case Study 3

Find the symbolic expression and the numerical value of the power absorbed by the current source in the circuit shown in Fig. 3.9a, with $E = 3V$, $A = 350\mu A$, $R_1 = 30k\Omega$, $R_2 = 20k\Omega$, $R_3 = 40k\Omega$.

From the right mesh (KVL) and from Ohm's law, we find the values for the R_3 descriptive variables. (See Fig. 3.9b.)

We have to compute the power absorbed by the current source, therefore we introduce as a further unknown the voltage v_A and, from the left mesh (KVL) and from Ohm's law, we find the values for the R_1 descriptive variables. (See Fig. 3.9b.) Then, from the KCL at node 1, we find the current in R_2 and (Ohm's law) the voltage across it. (See Fig. 3.9b.)

Fig. 3.9 Case Study 3

(a)

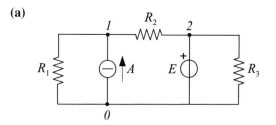

(b)

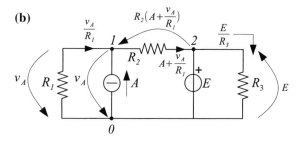

Finally, from the central mesh (KVL), we have $v_A + E + R_2(A + \frac{v_A}{R_1}) = 0$; that is, $v_A = -\frac{R_1}{R_1+R_2}(E + R_2 A) = -\frac{3}{5}(3 + 20 \cdot 10^3 \cdot 350 \cdot 10^{-6})\text{V} = -6$ V. Therefore the requested power is $p = v_A A = -6 \cdot 350 \cdot 10^{-6}$ W $= -2.1$ mW.

Case Study 4

Find the symbolic expression and the numerical value of the current i_E in the circuit shown in Fig. 3.10a, with $E = 1V$, $A = 1mA$, $R = 20k\Omega$.

Notice that this circuit corresponds in structure to the generic example used in Sect. 1.5.

We can start from the right mesh (KVL) and from Ohm's law to find the descriptive variables of resistor R in terms of the circuit parameters. (See Fig. 3.10b.)

Then, we can use the nodal KCL at node 3 and Ohm's law to find the descriptive variables of resistor $3R$ in terms of both the unknown i_E and the circuit parameters. (See Fig. 3.10c.)

Now we can find the descriptive variables of resistor $2R$ by using the KVL for the mesh defined by nodes 2–3–5 and Ohm's law. (See Fig. 3.10d)

Finally, using the nodal KCL at node 5, we have $i_E - \frac{E}{R} = \frac{E}{R} - \frac{3}{2}i_E + A$; that is, $i_E = \frac{4}{5}\frac{E}{R} + \frac{2}{5}A = 440\mu A$.

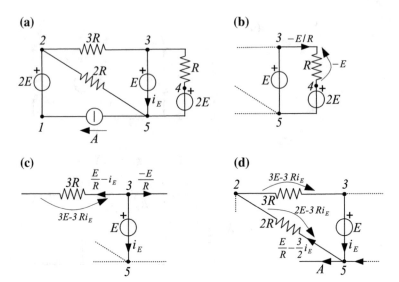

Fig. 3.10 Case Study 4

Case Study 5

Find the symbolic expression and the numerical value of the current i in the circuit shown in Fig. 3.11a, with $E = 22V$, $R = 10k\Omega$.

We can start from Ohm's law to find the voltage across resistor $R/2$. Now, we cannot go on without introducing a second unknown, that is, one of the descriptive variables of the other resistors. For instance, we choose the voltage (say v) across the upper left resistor. Then, we can exploit the KVL for mesh A and Ohm's law (twice), as shown in Fig. 3.11b. By using the KVL for mesh B, Ohm's law, the KCL at node 3, and Ohm's law again, we find the voltages and currents shown in Fig. 3.11c. Finally, using the nodal KCL at node 2 and KVL for mesh C (Fig. 3.11d), we have $i + \frac{v}{R} = \frac{E-v}{2R}$ and $E - v + \frac{Ri}{2} = \frac{v}{2} - \frac{5}{4}Ri$, respectively. After some algebra, we find $i = -\frac{2}{11}\frac{E}{R} = -\frac{4}{10^4}A = -400\mu A$.

You can check your comprehension by solving again the problem stated in Case Study 5, choosing a different second unknown.

3.2.5 Diode

A diode is an electronic two-terminal device (Fig. 3.12a shows two real devices). The symbol most commonly used for the model is shown in Fig. 3.12b: the arrow

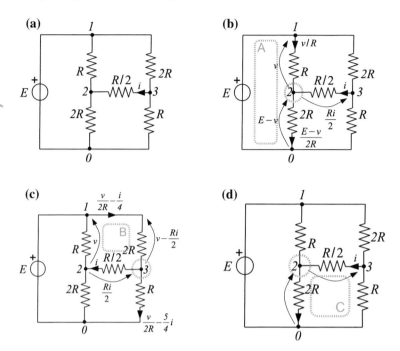

Fig. 3.11 Case Study 5

contained in the symbol suggests that, for a diode connected at two nodes of a circuit, the swapping of its terminals changes the electrical behavior. This asymmetry is typical of many components whose descriptive equation is nonlinear.

The **diode descriptive equation**[4] is

$$i(t) = I_S \left(e^{\frac{v(t)}{nV_t}} - 1 \right) \equiv I_S \psi(v; n) \qquad (3.6)$$

where I_S (with the physical dimension of ampere) is the reverse bias saturation current, V_t (with physical dimension of volt) is the thermal voltage, and n (dimensionless) is the ideality factor (or quality factor). The ideality factor n typically varies from 1 (case of an "ideal" diode) to 2 and accounts for imperfections in real

---

[4] Also called Shockley ideal diode equation, named after transistor coinventor William Bradford Shockley (1910–1989). With John Bardeen and Walter H. Brattain he invented the point contact transistor in 1947 and they were jointly awarded the 1956 Nobel Prize in Physics.

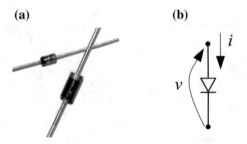

Fig. 3.12 Diode: **a** physical devices; **b** model

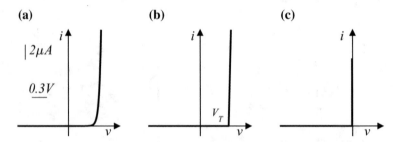

Fig. 3.13 Diode: **a** graphical representation of the descriptive equation (with $n = 1$, $V_t = 25.85$ mV, $I_S = 20$ nA); **b, c** PWL approximations. The two segments in panel **a** define the scales for the axes

devices. The descriptive equation is nonlinear and is represented on the (v, i) plane in Fig. 3.13a. It is evident that the diode has low resistance to the flow of current for large enough values of v, and high resistance for negative or slightly positive values of v. Panels b and c show two piecewise-linear (PWL) approximations of the original equation, often used for hand calculations. We remark that the voltages V_t (thermal voltage) and V_T (knee voltage) are different. In the case of panel b, the descriptive equation becomes:

$$i(t) = \begin{cases} 0 & \text{for } v < V_T \\ g(v - V_T) & \text{for } v \geq V_T \end{cases}.$$ (3.7)

In implicit form, it can be expressed as $f(i, v) = i - g\frac{v - V_T + |v - V_T|}{2} = 0$.
In the case of panel c, the descriptive equation is:

$$\begin{cases} i(t) = 0 & \text{for } v < 0 \\ v(t) = 0 & \text{for } i \geq 0 \end{cases}.$$ (3.8)

In implicit form, it can be expressed as $f(i, v) = vi = 0$, with $v \leq 0$ and $i \geq 0$.

The **driving-point (DP) characteristic**: for a two-terminal element such as those defined up to now, is the curve containing all admitted pairs on the plane (v, i).

For instance, the DP characteristics of a resistor and some alternatives for a diode are shown in Figs. 3.2c and 3.13, respectively. The asymmetric DP characteristic of the diode unveils its asymmetric behavior.

Case Study 1

Consider the descriptive equation 3.7 with the PWL characteristic shown in Fig. 3.13b. Show that this equation can be obtained as the descriptive equation of the composite two-terminal element shown in Fig. 3.14, where the diode's PWL characteristic is that represented in Fig. 3.14 (Eq. 3.8).

The first step to obtaining the characteristic is to write the topological equations:

$$\begin{cases} i_d = i \\ v = Ri + v_d + V_T \end{cases}. \tag{3.9}$$

Then we consider the two regions of the diode's characteristic, identified by an equation ($i_d = 0$ or $v_d = 0$) and by an inequality. Henceforth these regions are called α and β, respectively. (See Fig. 3.14) Combining descriptive equations and topological equations leads to the formulation of the PWL equations in terms of v and i:

- α region: taking $i_d = 0$, Eq. 3.9 gives $i = 0$ and $v = v_d + V_T$. This result holds as long as the inequality $v_d < 0$ is satisfied, that is, as long as $v - V_T < 0$. This corresponds to the first line of Eq. 3.7.
- β region: taking $v_d = 0$, we have $v = Ri + V_T$. This result holds as long as $i_d > 0$, for which we have $i_d(= i) = (v - V_T)/R > 0$; that is, $v > V_T$. For $R = 1/g$, the second line of Eq. 3.7 follows.

Fig. 3.14 Case Study 1

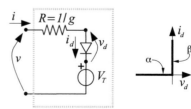

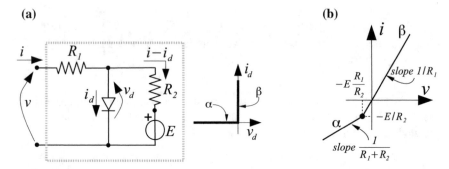

Fig. 3.15 **a** Composite two-terminal element (*left*) and diode's characteristic (*right*); **b** resulting PWL DP characteristic (for $E > 0$)

Case Study 2

Figure 3.15a *represents a two-terminal with descriptive variables i and v. Among the components internal to that element, the diode has the characteristic shown in the right part of the same figure. The voltage generator has* $E > 0$. *Under these assumptions, we want to get the two-terminal's descriptive equation and to represent the corresponding DP characteristic in the* (v, i) *plane.*

Following the track of Case Study 1, KCL and KVL easily lead us to write the following two equations, in which the v, i variables appear together with the descriptive variables of the diode.

$$\begin{cases} v_d = E + R_2(i - i_d) \\ v = R_1 i + v_d \end{cases} \qquad (3.10)$$

- Assuming that the diode is operating in the α region, so that $i_d = 0$ and $v_d < 0$, the previous equations easily give:

$$\begin{cases} E + R_2 i < 0 \\ v = E + (R_1 + R_2)i \end{cases} \qquad (3.11)$$

Thus inequality $v_d < 0$ is satisfied in the region $i < -E/R_2$ (i.e., $v < -E R_1/R_2$).
- Assuming now that the diode operates in the β region, the substitution $v_d = 0$ in Eq. 3.10 gives

$$\begin{cases} E + R_2(i - i_d) = 0 \\ v = R_1 i \end{cases} \qquad (3.12)$$

The $i_d \geq 0$ inequality is then satisfied for $i \geq -E/R_2$. The two branches of the obtained DP characteristic, indicated by α and β for obvious reasons, are shown in Fig. 3.15b.

3.2.6 Bipolar Junction Transistor

A bipolar junction transistor (or BJT) is a three-terminal device, whose symbol is shown in Fig. 3.16, that exploits two junctions between two semiconductor types (n-type and p-type) and uses both electron and hole charge carriers. In contrast, unipolar transistors, such as field-effect transistors (or FET), use only one kind of charge carrier.

Under proper assumptions, the **BJT descriptive equations**[5] are

$$\begin{cases} i_C = \alpha_F f_1(v_{EB}) - f_2(v_{CB}) \\ i_E = -f_1(v_{EB}) + \alpha_R f_2(v_{CB}) \end{cases}, \qquad (3.13)$$

where $f_1(v_{EB}) = I_{ES}\left(e^{\frac{-v_{EB}}{V_t}} - 1\right) = I_{ES}\psi(-v_{EB}; 1)$ and $f_2(v_{CB}) = I_{CS}$ $\left(e^{\frac{-v_{CB}}{V_t}} - 1\right) = I_{CS}\psi(-v_{CB}; 1)$, ψ is defined in Eq. 3.6, V_t is defined in Sect. 3.2.5, I_{ES}, I_{CS} (both with typical values ranging from 10^{-12} to 10^{-10} A at room temperature), α_F (typically ranging from 0.5 to 0.8), and α_R (usually set to 0.99) are device parameters.

Fig. 3.16 Symbol of the BJT

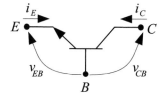

---

[5]This is the so-called Ebers-Moll model of the transistor, where the transistor is viewed as a pair of diodes. This model is named after Jewell James Ebers (1921–1959) and John Louis Moll (1921–2011), two American electrical engineers.

3.3   General Component Properties

On the basis of its descriptive equations, any component can be classified according to some criteria or properties. Here we introduce a set of basic properties.

3.3.1   Linearity

A component is **linear** if all its descriptive equations are linear and homogeneous with respect to the descriptive variables. Otherwise it is **nonlinear**.

For instance, the resistor is linear, whereas the ideal sources, the diode and the BJT are nonlinear.

Other examples:

- A two-terminal with descriptive equation $i = C\frac{dv}{dt}$ is linear, because the derivative (as well as the integral) is a linear operator.
- A three-terminal with descriptive equations

$$\begin{cases} v_1 = R(t)i_2 \\ i_1 = I_0 cos(\frac{v_2}{V_0}) + \alpha i_2 \end{cases} \tag{3.14}$$

is nonlinear, because at least one of the two equations (the second one) is nonlinear with respect to at least one descriptive variable. If the second equation were $i_1 = I_0 cos(\beta)\frac{v_2}{V_0} + \alpha i_2$, the three-terminal would be classified as linear, because in this case the term $\frac{I_0}{V_0}cos(\beta)$ plays the role of a mere coefficient.

3.3.2   Time-Invariance

A component is **time-invariant** if its descriptive equations do not depend on time. This means that the relationship between the descriptive variables or their derivatives (or integrals) is always the same, even if the values of the descriptive variables change in time, of course. Otherwise it is **time-varying**.

The resistor is time-invariant (remember that R is assumed to be constant) as well as the diode and the BJT, whereas the ideal sources are time-varying, in general, inasmuch as their descriptive equations contain a time-varying parameter (the impressed voltage/current).

Other examples:

- A two-terminal with descriptive equation $i = C \frac{dv}{dt}$ is time-invariant: i and v change in time, but the relationship between i and $\frac{dv}{dt}$ is always the same, because C is assumed to be constant.
- The three-terminal with descriptive equations 3.14 is time-varying, because at least one of the two equations (the first one) contains an explicit dependence on time for a parameter (the resistance R).
- A two-terminal with descriptive equation

$$
\begin{cases}
v = 0 & \text{for } t < t_0 \\
i = 0 & \text{for } t \geq t_0
\end{cases}
\tag{3.15}
$$

is time-varying, because the descriptive equation changes with time. In particular, this two-terminal is called an *ideal switch* and is closed (short circuit) for $t < t_0$ and open (open circuit) for $t \geq t_0$.

3.3.3 Memory

A component is **memoryless** or **resistive** or **adynamic** if its descriptive equations involve only the descriptive variables and not their derivatives (or integrals). Otherwise we say that it **has memory** or is **dynamic**.

Resistor, ideal sources, diode, and BJT are memoryless components.
Other examples are:

- A two-terminal with descriptive equation $i = C \frac{dv}{dt}$ has memory.
- The three-terminal with descriptive equations 3.14 is memoryless, because both equations are expressed in terms of the descriptive variables only, not involving their derivatives/integrals.

A circuit containing only linear, time-invariant, memoryless components and independent sources is called a **linear time-invariant resistive circuit**.

3.3.4 Basis

> A memoryless component admits the **voltage basis** (**current basis**) or is a
> **voltage-controlled resistor** (**current-controlled resistor**) if its descriptive
> equations can be expressed in the explicit form $i = f(v)$ $(v = g(i))$, where
> in general v is the vector of the descriptive voltages and i is the vector of the
> descriptive currents.

A two-terminal can admit the voltage basis, the current basis, both bases, or none
of them.

> An n-terminal element with $n > 2$ can also admit **mixed bases**: this means that
> at least an explicit form $y = f(x)$ is admitted, where the $n - 1$ independent
> variables x (i.e., the basis) are a mixed set of descriptive voltages and currents.

In other words, it must be possible to assign *arbitrarily* the vector x and obtain
univocally (through the vector of functions f) the $n - 1$ variables y. This restricts the
possible combinations because we cannot assign both voltage and current at a given
terminal. For instance, in a three-terminal, it is not possible to impose both v_1 and
i_1: if we impose v_1 with a voltage source, the current i_1 depends on the component
equations, and vice versa. (See also the examples below.) Then, for a three-terminal,
the possible mixed bases are (v_1, i_2) and (i_1, v_2), whereas (v_1, i_1) and (v_2, i_2) cannot
be bases.

For instance, the resistor admits both bases, the ideal voltage source and the short
circuit admit only the current basis, whereas the ideal current source and the open
circuit admit only the voltage basis. The diode with the characteristic of Fig. 3.13a
admits both bases, whereas in the PWL case of Fig. 3.13b it admits only the voltage
basis and in the case of Fig. 3.13c it does not admit any basis. The BJT admits all
possible bases.

Other examples are:

- A three-terminal with descriptive equations

$$\begin{cases} v_1 = Ri_2 \\ i_1 = I_0 cos(\frac{v_2}{V_0}) + \alpha i_2 \end{cases}$$

admits bases (i_1, v_2) and (v_1, v_2). (See Fig. 3.17.) It does not admit either the mixed
basis (v_1, i_2), because the two variables are proportional (then it is impossible to
assign both of them arbitrarily), or the current basis (i_1, i_2), because the $cos(\cdot)$
function is not bijective and then from the second equation it is not possible to
obtain univocally v_2 for a given pair (i_1, i_2);

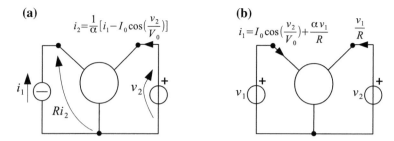

Fig. 3.17   Circuit verification of the existence of bases (i_1, v_2) **(a)** and (v_1, v_2) **(b)**

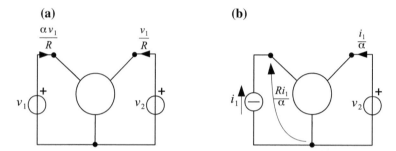

Fig. 3.18   Circuit verification of the existence of bases (v_1, v_2) **(a)** and (i_1, v_2) **(b)**

- A three-terminal with descriptive equations

$$\begin{cases} v_1 = Ri_2 \\ i_1 = \alpha i_2 \end{cases} \tag{3.16}$$

admits only the voltage basis and the mixed basis (i_1, v_2) (Fig. 3.18): the other two bases are not admitted due to the proportionality relationships imposed by the two descriptive equations.

- A three-terminal with descriptive equations

$$\begin{cases} v_1 = 0 \\ i_1 = \alpha i_2 \end{cases} \tag{3.17}$$

admits only the mixed basis (i_1, v_2) (Fig. 3.19): any basis containing v_1 is not admitted because (due to the first equation) v_1 cannot be arbitrarily assigned and the second equation makes it impossible to assign both currents arbitrarily.

Fig. 3.19 Circuit
verification of the existence
of basis (i_1, v_2)

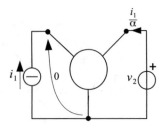

3.3.5 Energetic Behavior

The energetic behavior of a component depends on the expression of the absorbed
power p (for memoryless components) or energy w (for components with memory)
for any possible electrical situation, that is, for any admitted values of the descriptive
variables.

> A component is:
>
> - **Nonenergic** (or **inactive**) if $p(t) = 0 \, (w(t) = 0)$ for any electrical situation
> - **Passive** if $p(t) \geq 0 \, (w(t) \geq 0)$ for any electrical situation
> - **Strictly active** if $p(t) \leq 0 \, (w(t) \leq 0)$ for any electrical situation
> - **Active** if the sign of $p(t) \, (w(t))$ depends on the electrical situation

A memoryless component can be said to be **dissipative** as an alternative to passive.

For a given memoryless two-terminal element, the classification is quite easy on
the basis of its DP characteristic: if it lies only on the coordinate axes, the component
is nonenergic; if it lies only on the I–III (II–IV) quadrants, the component is passive
(strictly active); otherwise the component is active.

For instance, the resistor is passive (by assuming $R > 0$) because the absorbed
power is $p(t) = Ri^2 = v^2/R \geq 0$ for any electrical situation.

Instead, the ideal sources are active. By considering the voltage source, for
instance, the absorbed power is $p(t) = e(t)i(t)$ (Fig. 3.4a): because $i(t)$ depends
on the rest of the circuit, the sign of $p(t)$ depends on the specific electrical situation considered and nothing general can be stated. For instance, in Case Study 2 of
Sect. 3.2.4, one of the two voltage sources absorbs positive power, whereas the other
absorbs negative power.

The diode is passive, generally speaking, even if in the case of Fig. 3.13c it would
be nonenergic.

A resistor with negative resistance ($R < 0$) is strictly active.

The energetic behavior of components with memory is treated in Volume 2.

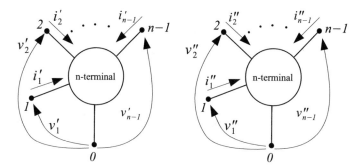

Fig. 3.20 The same n-terminal element in two different electrical situations

3.3.6 Reciprocity

The reciprocity property is encountered in various fields of physics, such as mechanics, acoustics, and electromagnetism. *In broad terms, it concerns the possible exchange of roles between input and output of a physical system.* In the case of electromagnetism this property was studied, among others, by Maxwell[6] [1] and by Lorentz[7] [2]. An example concerning electromagnetic propagation regards the possibility of swapping the positions of a transmitter and a receiver without altering the quality of the received signal. The reciprocity property is also present, however, in circuit theory, where each propagation phenomenon is absent by assumption.

One of the possible ways to begin the discussion about reciprocity within the circuit domain is to define this property for a memoryless n-terminal component. To this aim, we consider the same n-terminal element in two different electrical situations, for example, two identical components working in two different circuits, two identical components used in different places in the same circuit, or a single component working in a circuit but considered at two different times. In the first situation, we measure voltages v' and currents i', whereas in the second situation, we measure voltages v'' and currents i'', as shown in Fig. 3.20.

We can now introduce the (virtual) cross-powers:

$$p'(t) = (v'')^T i' = (i')^T v'' \tag{3.18}$$

$$p''(t) = (v')^T i'' = (i'')^T v' \tag{3.19}$$

---

[6]James Clerk Maxwell (1831–1879) was a Scottish scientist who formulated the classical theory of electromagnetic radiation, bringing together for the first time electricity, magnetism, and light as manifestations of the same phenomenon.

[7]Hendrik Antoon Lorentz (1853–1928) was a Dutch physicist who shared the 1902 Nobel Prize in Physics with Pieter Zeeman for the discovery and theoretical explanation of the Zeeman effect. He also derived the transformation equations that settled the basis for the special relativity theory of Albert Einstein.

An n-terminal is said to be **reciprocal** if $p'(t) = p''(t)$ for any pair of electrical situations, that is, for any v', v'', i', i'' compatible with the n-terminal. Otherwise it is nonreciprocal.

This definition may seem abstruse, far from the claims made at the beginning of this section about the links between an "input" and an "output" properly defined for the component. To show that this distance is only apparent, we consider a four-terminal, and assume that it is reciprocal and admits the bases shown in Fig. 3.21.

Referring to the first admitted basis, we first consider the four-terminal in the two electrical situations described in Fig. 3.22a and b. In both circuits, the current of terminal 3 is set to 0. In the first circuit, we assign the remaining pair $(v_1, v_2) = (E, 0)$. This is done through a voltage source E and a short circuit, as shown in Fig. 3.22a. The values of i_1', i_2', v_3' are obtained from the descriptive equations as functions of the components of the basis values $(E\ 0\ 0)^T$. Similar considerations apply to the second circuit, where $(v_1, v_2) = (0, E)$ and with i_1'', i_2'', v_3'' now calculated as functions of the basis values $(0\ E\ 0)^T$. The pairs of vectors v', i', v'', i'' associated with the two circuits:

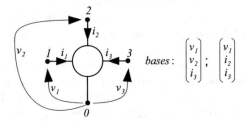

Fig. 3.21 The four-terminal element and its admitted bases

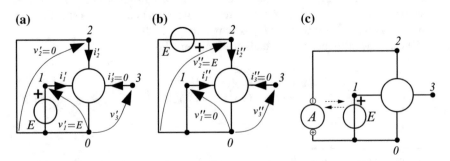

Fig. 3.22 First example

(a)   **(b)**   **(c)**

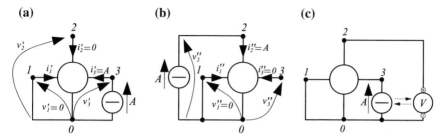

Fig. 3.23 Second example

$$v' = \begin{pmatrix} E \\ 0 \\ v'_3 \end{pmatrix}; \quad i' = \begin{pmatrix} i'_1 \\ i'_2 \\ 0 \end{pmatrix}; \quad v'' = \begin{pmatrix} 0 \\ E \\ v''_3 \end{pmatrix}; \quad i'' = \begin{pmatrix} i''_1 \\ i''_2 \\ 0 \end{pmatrix}.$$

can now be used for the calculation of the virtual cross-powers $(v')^T i''$ and $(v'')^T i'$. By equating these terms, as required by the reciprocity assumption, we get $i''_1 = i'_2$, that is, the current of terminal 1 in the second circuit equals that of terminal 2 in the first circuit. In other words, in the circuit of Fig. 3.22c, the source E (input) and the ammeter (output) can be swapped without changes of the ammeter's measurement.

As a second example, referring now to the second basis, we consider $v_1 = 0$ and we set $(i_2, i_3) = (0, A)$ in Fig. 3.23a and $(i_2, i_3) = (A, 0)$ in Fig. 3.23b. We obtain the following two pairs of vectors for the descriptive variables.

$$v' = \begin{pmatrix} 0 \\ v'_2 \\ v'_3 \end{pmatrix}; \quad i' = \begin{pmatrix} i'_1 \\ 0 \\ A \end{pmatrix}; \quad v'' = \begin{pmatrix} 0 \\ v''_2 \\ v''_3 \end{pmatrix}; \quad i'' = \begin{pmatrix} i''_1 \\ A \\ 0 \end{pmatrix}$$

The reciprocity condition provides $v'_2 = v''_3$. Therefore, in the circuit of Fig. 3.23c the swapping of the current source (input) and the voltmeter (output) does not change the voltmeter's measurement.

To check your comprehension, you can refer to the first basis and try to obtain the results when v_2 (instead of v_1) or i_3 are set to zero.

Both previous results bring back the reciprocity to a swapping between a source and a measuring instrument. We can have a third reciprocity relation that does not refer to a swapping.

To show this, we refer again to the first basis, taking $v_1 = 0$ and working on the remaining pair of basis variables, v_2 and i_3. Therefore in the circuit of Fig. 3.24a we set $i'_3 = A$ and $v'_2 = 0$, whereas in the circuit of Fig. 3.24b (identical to Fig. 3.22b, but repeated for clarity) we set $i''_3 = 0$ and $v''_2 = E$. This amounts to having these two pairs of vectors of descriptive variables:

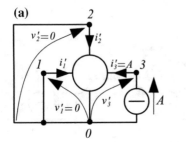 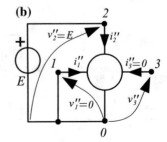

Fig. 3.24 Third example

$$
v' = \begin{pmatrix} 0 \\ 0 \\ v_3' \end{pmatrix} ; \quad i' = \begin{pmatrix} i_1' \\ i_2' \\ A \end{pmatrix} ; \quad v'' = \begin{pmatrix} 0 \\ E \\ v_3'' \end{pmatrix} ; \quad i'' = \begin{pmatrix} i_1'' \\ i_2'' \\ 0 \end{pmatrix}
$$

in which the values of $v_3', i_1', i_2', v_3'', i_1'', i_2''$ are obtained by the descriptive equations. Now, writing the reciprocity condition $(v')^T i'' = (v'')^T i'$ we obtain $E i_2' + A v_3'' = 0$ or $i_2' = -\frac{A}{E} v_3''$.

To check your comprehension, you can try to obtain the analogous result when v_2, instead of v_1, is taken fixed to zero.

Coming back to the general case, for the introduced two-terminal elements, we can check reciprocity by applying the proposed definition to their constitutive equations.

For instance, a resistor (with $0 < R < \infty$) admits both bases. With this in mind, it is very easy to verify that it is reciprocal. Indeed, for two pairs (v', i') and (v'', i'') of values of the resistor's descriptive variables, we can always write $v'i'' = Ri'i''; v''i' = Ri''i'$. Thus, $p' = p''$ always.

A diode is nonreciprocal because of its nonlinear DP characteristic. To show this, we first denote by $i = f(v)$ the diode's descriptive equation in the case of both Eqs. 3.6 and 3.7. Then we observe that, for two assigned voltage values v' and v'', we have $i' = f(v')$ and $i'' = f(v'')$, respectively. Therefore, because the corresponding cross-power terms are $v'i'' = v'f(v'')$ and $v''i' = v''f(v')$, we conclude that $p' \neq p''$. With similar reasoning, you can check that the BJT is also nonreciprocal.

Voltage and current sources are nonreciprocal components as well. A voltage source e, for instance, admits the current basis, and for two assigned current values i' and i'' (with $i' \neq i''$), we obviously have $v' = v'' = e$. Thus we conclude that $p' \neq p''$.

Remark For dynamic components, reciprocity can be defined by making reference to virtual works instead of virtual cross-powers.

Finally, we introduce a theorem here that is proved in Sect. 6.3.

Theorem 3.1 (Reciprocity theorem) *A component consisting of reciprocal elements only is in turn reciprocal.*

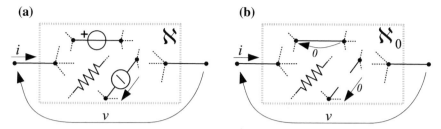

Fig. 3.25 A composite two-terminal ℵ (**a**) and the corresponding two-terminal ℵ₀, obtained from ℵ by turning off all the independent sources (**b**)

This is a remarkable result that allows one easily to check the reciprocity of some composite components. For instance, owing to this theorem, we can immediately assess that any n-terminal component containing only resistors is reciprocal. We remark that the theorem provides a *sufficient* condition only. This means that a composite n-terminal containing both reciprocal and nonreciprocal components can be either reciprocal or nonreciprocal: we have to check it through the reciprocity definition.

3.4 Thévenin and Norton Equivalent Representations of Two-Terminal Resistive Components

A generic (black box) two-terminal ℵ made up of linear, time-invariant, and memoryless components and independent sources (Fig. 3.25a) might admit an equivalent representation with the same descriptive equation but a simplified internal structure, which neglects the detail of all internal variables. This is sometimes called a *macromodel* of the two-terminal. There are two possible circuit representations for macromodels, known as the Thévenin equivalent and Norton equivalent. In practice, a macromodel is a higher-level model (Sect. 1.1) of a portion of circuit.

3.4.1 Thévenin Equivalent

The Thévenin equivalent is shown in Fig. 3.26a.[8]

---

[8]The original theorem (commonly known as Thévenin's theorem) was independently derived in 1853 by the German scientist Hermann von Helmholtz [3] and in 1883 by the French electrical engineer Léon Charles Thévenin (1857–1926) [4, 5].

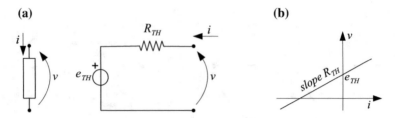

Fig. 3.26 Thévenin equivalent (**a**) and corresponding DP characteristic (**b**)

The descriptive equation of this two-terminal is

$$v(t) = e_{TH}(t) + R_{TH}i(t) \tag{3.20}$$

which corresponds to the DP characteristic shown in Fig. 3.26b. This equation states that this representation is admitted by the two-terminal ℵ when its constitutive equation can be recast in the explicit form Eq. 3.20, that is, *if and only if it admits the current basis*.

How do we find the two parameters e_{TH} and R_{TH} for the Thévenin equivalent of ℵ? We have two ways:

1. One can find the descriptive equation of ℵ, compare it with Eq. 3.20, and obtain e_{TH} and R_{TH} by comparison. This can be easily done for elementary circuits, but becomes harder for more complex circuits.
2. Otherwise, one can analyze two simpler two-terminals: the first one (say ℵ$_{oc}$) is obtained from ℵ by imposing $i = 0$ and allows obtaining the open-circuit voltage e_{TH} (Eq. 3.20); the second one (say ℵ$_0$; see Fig. 3.25b) is obtained from the original one by turning off all the independent sources (i.e., by replacing each voltage source with a short circuit and each current source with an open circuit) and allows obtaining the Thévenin equivalent resistance as $R_{TH} = \frac{v}{i}$. Indeed (Eq. 3.20), R_{TH} is the resistance of ℵ when it is equivalent to a linear resistor (then passive), that is, when all the independent sources are turned off, thus ensuring that $e_{TH} = 0$.

Remark: Equation 3.20 holds for the Thévenin equivalent shown in Fig. 3.26. If the voltage source is assumed upside down, there is just a change of sign in Eq. 3.20 but we follow the same line of reasoning. This is apparent in Case Study 2 below.

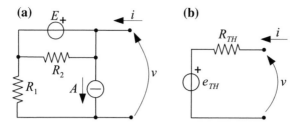

Fig. 3.27 Composite two-terminal ℵ for Case Study 1

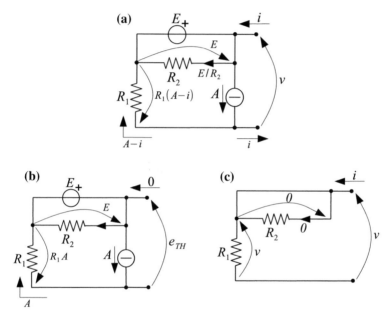

Fig. 3.28 Solution of Case Study 1: **a** Way 1; auxiliary two-terminals ℵ$_{oc}$ (**b**) and ℵ$_0$ (**c**) used in Way 2

Case Study 1

Find the Thévenin equivalent of the composite two-terminal shown in Fig. 3.27.

As stated above, we can solve this problem in two different ways.

Way 1. We find the descriptive equation of the two-terminal.

It is easy to find currents and voltages on the circuit by using Kirchhoff's laws and descriptive equations of the components, as shown in Fig. 3.28a.

Then, from the outer loop, we have the descriptive equation (to be compared with that of the Thévenin equivalent) $v = E - R_1(A - i) = \underbrace{E - R_1 A}_{e_{TH}} + \underbrace{R_1}_{R_{TH}} i$.

Way 2. We analyze two simpler auxiliary two-terminals.

The two-terminal $\aleph_{oc}$ (with $i = 0$ and $v = e_{TH}$) is shown in Fig. 3.28b and can be easily solved, thus finding $e_{TH} = E - R_1 A$.

The second auxiliary two-terminal $\aleph_0$ (with independent sources turned off) is shown in Fig. 3.28c and provides $R_{TH} = \frac{v}{i} = R_1$.

As stated above, the second way is more suitable as far as the circuit becomes more complex.

Notice that R_2 is not involved in the solution. Why?

Case Study 2

Find the Thévenin equivalent of the composite two-terminal $\aleph$ shown in Fig. 3.29a.

Notice that in this case the voltage source of the Thévenin equivalent is oriented differently than usual and corresponds to a descriptive equation $v(t) = -e_{TH}(t) + R_{TH}i(t)$. This means that in this case $e_{TH}(t) = -v\big|_{i=0}$.

The solution is obtained according to the second solving method. The solution according to the first method is left to the reader.

The auxiliary two-terminal $\aleph_{oc}$ (with $i = 0$ and $v = -e_{TH}$) is shown in Fig. 3.29b and can be easily solved, thus finding $e_{TH} = RA - 2E$.

The second auxiliary two-terminal $\aleph_0$ (with independent sources turned off) is shown in Fig. 3.29c and provides $R_{TH} = \frac{v}{i} = R$.

Notice that the left part of the two-terminal in Fig. 3.29a is not involved in the solution. Why?

To check your comprehension, you can try to solve Case Study 5 in Sect. 3.2.4 by preliminarily finding the Thévenin equivalent of the composite two-terminal connected to the resistor where the unknown i flows and then by solving the very simple resulting circuit.

3.4.2   Norton Equivalent

The Norton equivalent is shown in Fig. 3.30a.[9]

---

[9]The original theorem (commonly known as Norton's theorem) was independently derived in 1926 by Siemens and Halske researcher Hans Ferdinand Mayer (1895–1980) [6] and Bell Labs engineer Edward Lawry Norton (1898–1983) [7].

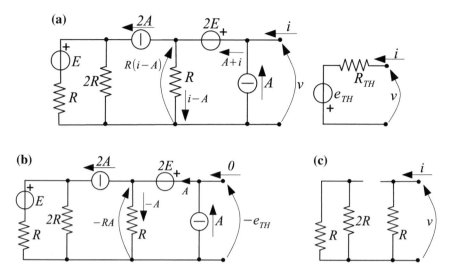

Fig. 3.29 a Two-terminal ℵ for Case Study 2; **b, c** auxiliary two-terminals

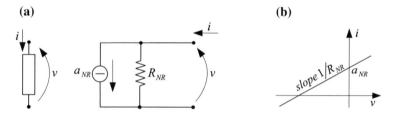

Fig. 3.30 Norton equivalent

The descriptive equation of this two-terminal is

$$i(t) = a_{NR}(t) + \frac{v(t)}{R_{NR}} \tag{3.21}$$

which corresponds to the DP characteristic shown in Fig. 3.30b. This equation states that this representation is admitted by the two-terminal when its constitutive equation can be recast in the explicit form 3.21, that is, *if and only if it admits the voltage basis*.

How do we find the two parameters a_{NR} and R_{NR}? Also in this case, we have a double key to uncover the Norton equivalent of a given generic two-terminal:

Fig. 3.31 First auxiliary two-terminal $\aleph_{sc}$ for Case Study 1

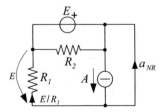

1. One can find the descriptive equation of the two-terminal $\aleph$ and obtain a_{NR} and R_{NR} by comparison with Eq. 3.21. This can easily be done for elementary circuits, but becomes harder for more complex circuits.
2. Otherwise, one can analyze two simpler two-terminals: the first one (say $\aleph_{sc}$) is obtained from $\aleph$ by imposing $v = 0$ and allows us to obtain the short-circuit current a_{NR} (Eq. 3.21); the second one is once more $\aleph_0$ and allows us to obtain the Norton equivalent resistance from $\dfrac{1}{R_{NR}} = \dfrac{i}{v}$.[10] Indeed (Eq. 3.21), R_{NR} is the resistance of the two-terminal $\aleph$ when it is equivalent to a linear resistor (then passive), that is, when all the independent sources are turned off, thus ensuring that $a_{NR} = 0$.

Remark: Equation 3.21 holds for the Norton equivalent shown in Fig. 3.30. If the current source is connected upside down, there is just a change of sign in Eq. 3.21 but we follow the same line of reasoning. This is apparent in Case Study 2 below.

Case Study 1

Find the Norton equivalent of the composite two-terminal shown in Fig. 3.27.

As stated above, we can solve this problem in two different ways.
Way 1. From the descriptive equation of the two-terminal, we easily find

$$i = \frac{v}{R_1} + A - \frac{E}{R_1} = \underbrace{A - \frac{E}{R_1}}_{a_{NR}} + \underbrace{\frac{1}{R_1}}_{\frac{1}{R_{NR}}} v.$$

Way 2. We analyze two simpler auxiliary two-terminals.
$\aleph_{sc}$ (with $v = 0$ and $i = a_{NR}$) is shown in Fig. 3.31 and can be easily solved, thus finding $a_{NR} = A - \frac{E}{R_1}$.
$\aleph_0$ (with independent sources turned off) was shown in Fig. 3.28c and provides $R_{NR} = R_{TH} = \frac{v}{i} = R_1$.

---

[10]We point out that R_{NR} can tend to infinite (in which case the Norton equivalent is just a current source), but cannot be null, because the two-terminal $\aleph$ admits the voltage basis by assumption.

Fig. 3.32 Norton equivalent
for Case Study 2

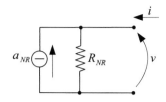

Fig. 3.33 Auxiliary
two-terminal $\aleph_{sc}$ for Case
Study 2

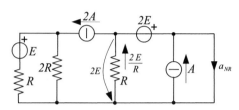

Case Study 2

Find the Norton equivalent shown in Fig. 3.32 of the composite two-terminal of Fig. 3.29a.

Notice that in this case the current source of the Norton equivalent is oriented differently than usual and corresponds to a descriptive equation $i(t) = -a_{NR}(t) + \frac{v(t)}{R_{NR}}$. This means that in this case $a_{NR}(t) = -i\big|_{v=0}$.

The solution is obtained according to the second solving method. The solution according to the first method is left to the reader.

$\aleph_{sc}$ (with $v = 0$ and $i = -a_{NR}$) is shown in Fig. 3.33 and can be easily solved, thus finding $a_{NR} = \frac{2E}{R} - A$.

$\aleph_0$ (with independent sources turned off) was shown in Fig. 3.29c and provides $R_{NR} = R_{TH} = \frac{v}{i} = R$.

3.4.3 Comparisons Between the Two Equivalent Models

It is quite apparent from the proposed examples that when a composite two-terminal admits both bases (and then both equivalent representations), it is straightforward to derive one model from the other. For instance, starting from Eq. 3.21, one easily finds $v = -R_{NR} \cdot a_{NR}(t) + R_{NR} \cdot i$. By comparing this equation with Eq. 3.20, we identify $e_{TH} = -R_{NR} \cdot a_{NR}$ and $R_{TH} = R_{NR}$.

Why is there such a large temporal interval (1853 vs. 1926) between the formulations of these seemingly similar equivalent models? The current-source equivalent did not occur to early electrical scientists because of the apparent impossibility of the existence of a current source. Only later did Norton and Mayer realize that the current-source equivalent was easier to use in certain practical situations.

3.5   Series and Parallel Connections of Two-Terminals

When analyzing a circuit, it is useful to replace some circuit parts whose details are not under study with others, equivalent but simpler. In other words, once more we replace some parts with macromodels. The most common structures admitting simplifications are series and parallel connections of two-terminals.

3.5.1   Series Connection

As shown in Fig. 3.34,

> Two two-terminals are connected **in series** when the same current flows through them.

The two-terminals a and b share the same current, whereas $v = v_a + v_b$.

For memoryless components, this connection makes sense if at least one of the two two-terminals admits the current basis.

3.5.1.1   Examples When both Components Admit the Current Basis

Figure 3.35a shows two resistors connected in series. In this case the descriptive equation of the composite two-terminal is $v = (R_A + R_B)i$, which is the descriptive equation of a resistor with resistance $R_A + R_B$. Then, if we are not interested in determining v_A or v_B, we can replace this series connection with the single resistor shown in Fig. 3.35b.

Figure 3.36a shows two voltage sources connected in series. In this case the descriptive equation of the composite two-terminal is $v = E_A - E_B$, which is the descriptive equation of a voltage source with impressed voltage $E_A - E_B$. Then we

Fig. 3.34 Two two-terminals connected in series

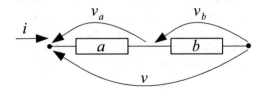

Fig. 3.35 Two resistors connected in series (**a**) and their equivalent model (**b**)

Fig. 3.36 Two voltage sources connected in series (**a**) and their equivalent model (**b**)

Fig. 3.37 Two components connected in series (**a**) and their equivalent model (**b**)

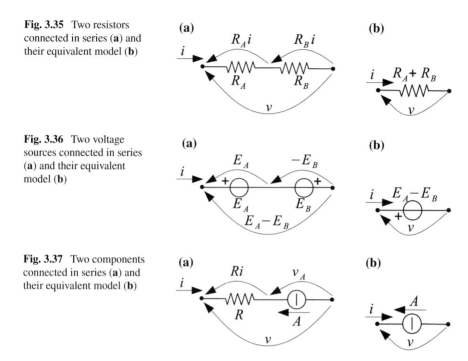

can replace this series connection with the single voltage source shown in Fig. 3.36b.

3.5.1.2 Examples When only One Component Admits the Current Basis

Figure 3.37a shows a resistor connected in series with a current source, which does not admit the current basis. In this case the descriptive equation of the composite two-terminal is $i = -A$, which is the descriptive equation of a current source. Then, if we are not interested in determining v_A or v_R, we can replace this series connection with the single current source shown in Fig. 3.37b (equivalent model). Notice that the voltage across the equivalent model is different from the voltage across the original current source.

Figure 3.38a shows a voltage source connected in series with a current source. The descriptive equation of the composite two-terminal is $i = -A$, which is the descriptive equation of a current source with impressed current $-A$. Then we can replace this series connection with the single current source shown in Fig. 3.38b.

Fig. 3.38 Two components connected in series (**a**) and their equivalent model (**b**)

Fig. 3.39 Two current sources connected in series (absurd connection)

Fig. 3.40 Two open circuits connected in series (undetermined connection)

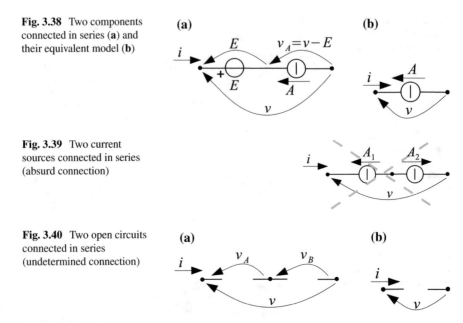

In any case, the equivalent model coincides (apart from the voltage) with the two-terminal not admitting the current basis.

3.5.1.3 Other Examples

When none of the two-terminals admits the current basis, we can either have an absurd or an undetermined situation. Figure 3.39 shows two current sources connected in series. For $A_1 \neq -A_2$, the KCL is violated. Then, this connection is absurd.

Figure 3.40a shows two open circuits connected in series. In this case there is no way to determine v_A and v_B. An equivalent model exists anyway. (See Fig. 3.40b.)

Finally, a short circuit connected in series with a generic two-terminal is obviously equivalent to the two-terminal itself.

3.5.2 Parallel Connection

As shown in Fig. 3.41,

Two two-terminals are **in parallel** when they are connected to the same pair of nodes and they share the same descriptive voltage.

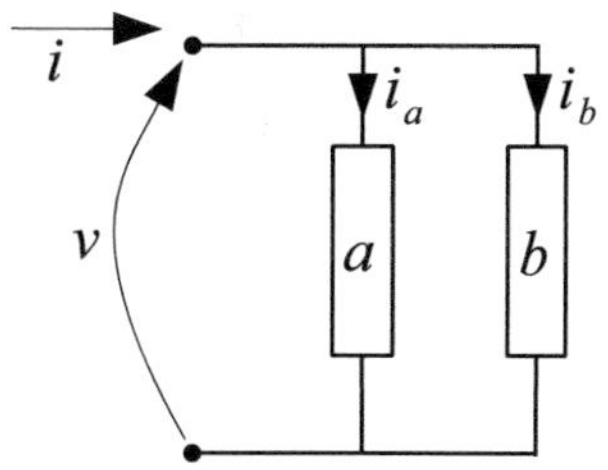

Fig. 3.41 Two
two-terminals connected in
parallel

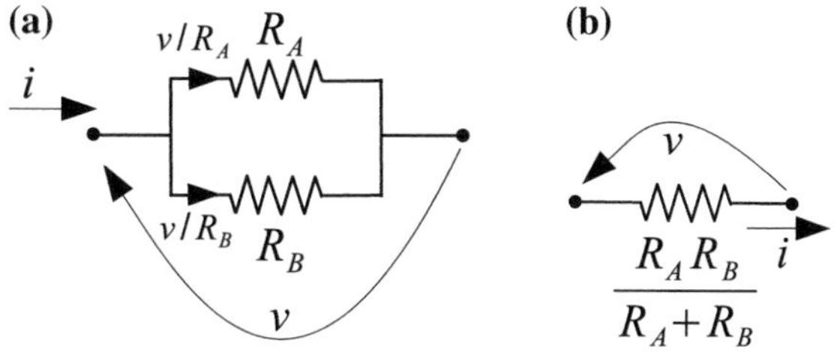

Fig. 3.42 Two resistors
connected in parallel (**a**) and
their equivalent model (**b**)

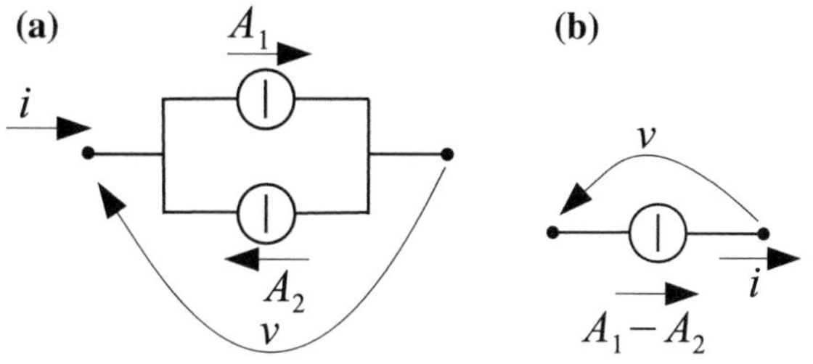

Fig. 3.43 Two current
sources connected in parallel
(**a**) and their equivalent
model (**b**)

The two-terminals a and b share the same voltage, whereas $i = i_a + i_b$.

For memoryless components, this connection makes sense if at least one of the two two-terminals admits the voltage basis.

3.5.2.1 Examples When both Components Admit the Voltage Basis

Figure 3.42a shows two resistors connected in parallel. In this case the descriptive equation of the composite two-terminal is $i = \frac{v}{R_A} + \frac{v}{R_B}$, which is the descriptive equation of a resistor with resistance $\frac{R_A R_B}{R_A + R_B}$. Thus if we are not interested in determining i_A or i_B, we can replace this parallel connection with the single resistor shown in Fig. 3.42b.

Figure 3.43a shows two current sources connected in parallel. In this case the descriptive equation of the composite two-terminal is $i = A_1 - A_2$, which is the descriptive equation of a current source with impressed current $A_1 - A_2$. Thus we can replace this parallel connection with the single current source shown in Fig. 3.43b.

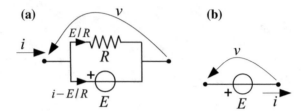

Fig. 3.44 Two components connected in parallel (**a**) and their equivalent model (**b**)

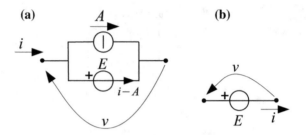

Fig. 3.45 Two components connected in parallel (**a**) and their equivalent model (**b**)

3.5.2.2   Examples When only One Component Admits the Voltage Basis

Figure 3.44a shows a resistor connected in parallel with a voltage source, which does not admit the voltage basis. In this case the descriptive equation of the composite two-terminal is $v = E$, which is the descriptive equation of a voltage source. Thus if we are not interested in determining the currents in the two branches, we can replace this parallel connection with the single voltage source shown in Fig. 3.37b (equivalent model). Notice that the current in the equivalent model is different from the current in the original voltage source.

Figure 3.45a shows a voltage source connected in parallel with a current source. The descriptive equation of the composite two-terminal is once more $v = E$, which is the descriptive equation of a voltage source with impressed voltage E. Thus also in this case we can replace this parallel connection with the single voltage source shown in Fig. 3.45b.

In any case, the equivalent model coincides (apart from the current) with the two-terminal not admitting the voltage basis.

3.5.2.3   Other Examples

When none of the two-terminals admits the voltage basis, we can either have an absurd or an undetermined situation. Figure 3.46a shows a voltage source and a short circuit connected in parallel. In this case the KVL is violated, therefore this connection is

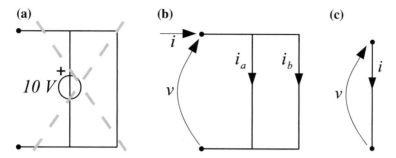

Fig. 3.46 a A voltage source and a short circuit connected in parallel (absurd connection); **b** two short-circuit sources connected in parallel (undetermined connection); **c** two-terminal equivalent to **b**

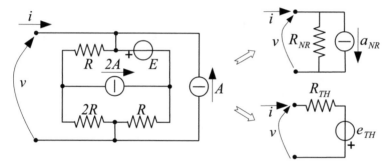

Fig. 3.47   Case Study

Fig. 3.48   Thévenin auxiliary two-terminal $\aleph_0$

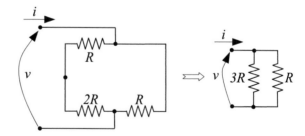

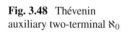

absurd. For the same reason, a parallel connection of two voltage sources impressing different voltages is absurd.

Figure 3.46b shows two short circuits connected in parallel. In this case there is no way to determine i_a and i_b, but an equivalent circuit exists. (See Fig. 3.46c.)

Finally, an open circuit connected in parallel with a generic two-terminal is obviously equivalent to the two-terminal itself.

Case Study

Find the Thévenin and Norton equivalents of the composite two-terminal shown in Fig. 3.47.

The solution is obtained according to the second solving method. It is left to the reader to find the solution according to the first method.

The auxiliary two-terminal $\aleph_0$ with independent sources turned off is shown in Fig. 3.48. The left resistors are connected in series and can be replaced by a single resistor with resistance $3R$. Then, $3R$ and R are in parallel, so that $R_{TH} = R_{NR} = \frac{3}{4}R$.

The Thévenin auxiliary two-terminal $\aleph_{oc}$ with $i = 0$ and $v = -e_{TH}$ is shown in Fig. 3.49. Owing to KVL for mesh A and to Ohm's law, we find the voltage and current for the bottom right resistor. Then, by using KCL at node 0 and Ohm's law, the voltage and current for the bottom left resistor can also be easily found. As a further step, KCL at node 1 and Ohm's law provide voltage and current also for the top left resistor. Finally, from KVL for the left mesh we have $-e_{TH} = 3RA + E + e_{TH} + 2RA + 2E + 2e_{TH}$; that is, $e_{TH} = -\frac{5}{4}RA - \frac{3}{4}E$.

The Norton auxiliary two-terminal $\aleph_{sc}$ with $v = 0$ and $i = a_{NR}$ is shown in Fig. 3.50. The circled numbers denote a possible sequence of steps one can take to find (from the KVL for the left mesh) finally the solution $a_{NR} = -\frac{5}{3}A - \frac{E}{R}$.

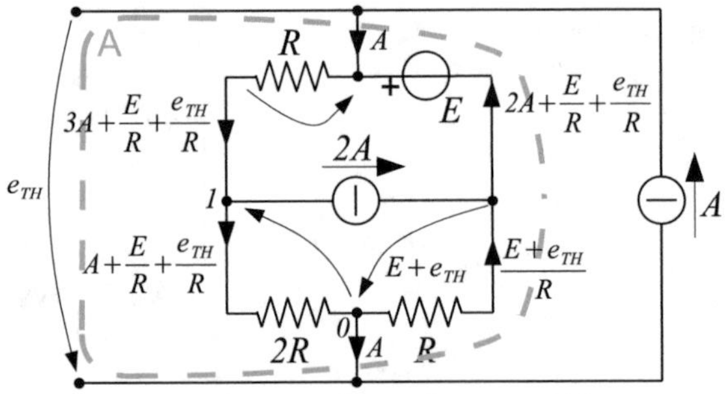

Fig. 3.49 Thévenin auxiliary two-terminal $\aleph_{oc}$

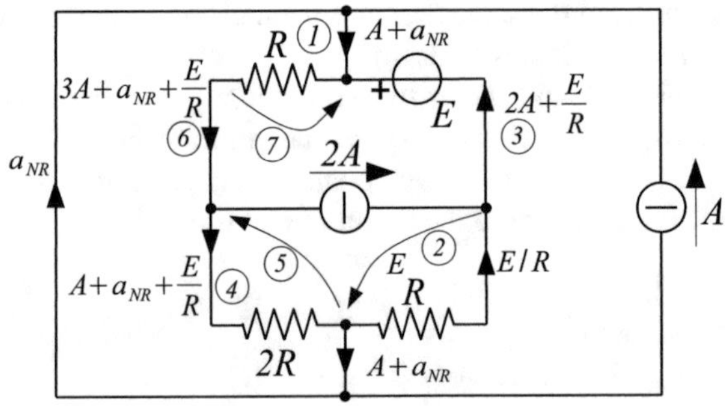

Fig. 3.50 Norton auxiliary two-terminal $\aleph_{sc}$

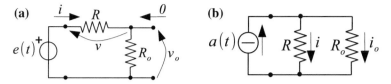

Fig. 3.51 a Voltage divider; **b** current divider

3.5.3 Numerical Aspects

When we have two resistors with resistances differing by orders of magnitude,

- If they are connected in series, the whole connection is practically equivalent to the resistor with the highest resistance value.
- If they are connected in parallel, the whole connection is practically equivalent to the resistor with the lowest resistance value.

For instance, if we consider two resistors with resistances $R_1 = 100\Omega$ and $R_2 = 1M\Omega$, their series connection is equivalent to a resistor with resistance $R_s = R_1 + R_2 \approx R_2$, whereas their parallel connection is equivalent to a resistor with resistance $R_p = \dfrac{R_1 R_2}{R_1 + R_2} \approx R_1$.

These "rules of thumb" can be easily generalized to the cases with more than two resistors connected either in series or in parallel.

3.6 Resistive Voltage and Current Dividers

In this section, we introduce two very common circuit structures: the voltage divider and the current divider.

3.6.1 Resistive Voltage Divider

The resistive voltage divider is shown in Fig. 3.51a. We remark that **the two resistors are connected in series**. We want to know which fraction of the input voltage $e(t)$ drops on the output resistor R_o. Because the two resistors are connected in series, we have $i = \dfrac{e(t)}{R + R_o}$. Then $v_o = e(t)\dfrac{R_o}{R + R_o}$.

Inasmuch as the current is the same in both resistors and due to the linear descriptive equation of the resistor, this equation simply means that the higher R_o is, the higher the fraction of $e(t)$ which drops on R_o.

3.6.2   Resistive Current Divider

The resistive current divider is shown in Fig. 3.51b. We remark that **the two resistors are connected in parallel**. We want to know which fraction of the input current $a(t)$ flows in the output resistor R_o. Because the two resistors are connected in parallel, we have $v = a(t) \dfrac{R R_o}{R + R_o}$. Then $i_o = a(t) \dfrac{R}{R + R_o}$.

Inasmuch as the voltage is the same across both resistors and due to the linear descriptive equation of the resistor, this equation simply means that the higher R_o is, the lower the fraction of $a(t)$ which flows in R_o. This is not surprising: the current tends to privilege the path with minimum resistance! A similar reasoning could be applied to a liquid flowing in two parallel pipes: the higher quantity of liquid will flow in the pipe with larger diameter (i.e., lower resistance).

3.7   Problems

3.1 The three-terminal shown in Fig. 3.52a has descriptive equations

$$\begin{cases} v_1 = R i_1 \\ v_2 = R_1 i_1 + R_2 i_2 \end{cases}$$

with $R, R_1, R_2 > 0$. Find the descriptive equations in terms of the descriptive variables shown in Fig. 3.52b.

3.2 Check the properties of linearity, time-invariance, and memory of the three-terminal shown in Fig. 3.52a for the following sets of descriptive equations, by assigning to the parameters $\alpha, \beta, \gamma, \delta,$ and σ proper physical dimensions, case by case.

(a) $\begin{cases} v_1 = \beta i_1 + \gamma i_2 \\ i_1 = \delta \frac{dv_2}{dt} \end{cases}$ (b) $\begin{cases} \alpha v_1^3 = i_1 + \gamma i_2^2 \\ v_1 = \delta \frac{di_2}{dt} \end{cases}$ (c) $\begin{cases} v_1 = \beta i_1 + \gamma i_2 \\ v_2 = \sigma i_2 \end{cases}$ (d) $\begin{cases} v_1 = \beta \sin\left(\frac{i_2}{I_0}\right) \\ \gamma v_2 = i_1 + \sigma i_2 \end{cases}$

3.3 For each composite two-terminal shown in Fig. 3.53, find:

1. Descriptive equation
2. Admitted bases
3. Energetic behavior
4. Thévenin and Norton equivalent representations shown in Figs. 3.26a and 3.30a, respectively (if admitted).

3.4 Classify each composite two-terminal shown in Fig. 3.54 from the energetic standpoint, assuming the following descriptive equations for the nonlinear two-terminals: $i_2 = \alpha v_2^2$ and $i_1 = \beta e^{\frac{v_1}{V_0}}$, with $\alpha = 1\frac{A}{V^2}$, $\beta = 1A$, $V_0 = 1V$, $E > 0$, $A > 0$.

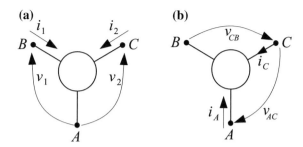

Fig. 3.52 Problems 3.1 and 3.2

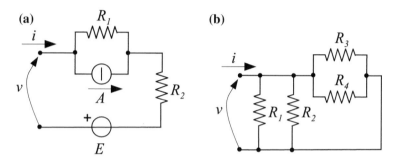

Fig. 3.53 Problem 3.3

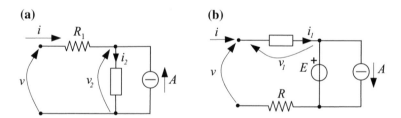

Fig. 3.54 Problem 3.4

3.5 Find, whenever possible, the Thévenin and Norton equivalent representations (shown in Fig. 3.26a and 3.30a, resp.) for the two-terminals shown in Fig. 3.55.

3.6 Find the expression and the value of the current i for the circuit shown in Fig. 3.56a, where $R_1 = R_3 = 1k\Omega$, $R_2 = 3k\Omega$, $E_1 = 1V$, $E_2 = 4V$.

3.7 For the circuit shown in Fig. 3.56b, where the values of A and E can be either positive or negative, find:

1. i.
2. v.
3. Power absorbed by the current source.
4. Power absorbed by the voltage source.

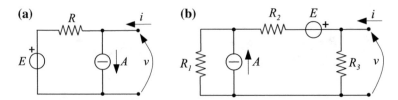

Fig. 3.55 Problem 3.5

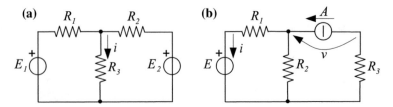

Fig. 3.56 a Problem 3.6. **b** Problem 3.7

Fig. 3.57 Problem 3.8

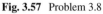

Fig. 3.58 a Problem 3.9. **b**
Problem 3.10

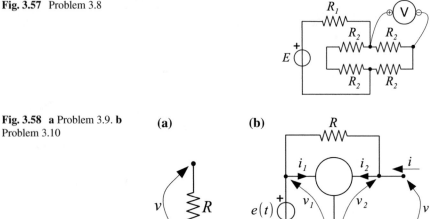

5. Is it possible that the latter two are both positive?

3.8 For the circuit shown in Fig. 3.57, find the expression of R_1 such that the volt-meter measures a voltage of $\frac{E}{4}$.

3.9 Choose the right answer(s). For the resistor shown in Fig. 3.58a, we can state that the absorbed power:

1. Is always ≥ 0

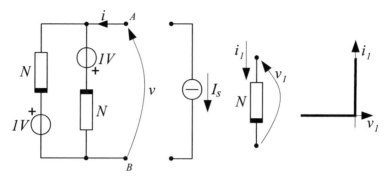

Fig. 3.59 Problem 3.11

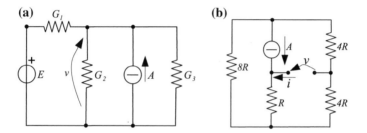

Fig. 3.60 a Problem 3.12. **b** Problem 3.13

2. Is always ≤ 0
3. Has a sign that depends on the values of v and i

3.10 Find the Norton equivalent representation (shown in Fig. 3.30a) for the two-terminal shown in Fig. 3.58b, where the three-terminal's descriptive equations are $i_1 = g_1 v_1 + g_M v_2$ and $i_2 = g_2 v_2$, with $g_2 \neq -1/R$.

3.11 By assuming that the nonlinear component N shown in Fig. 3.59 is a diode with the DP characteristic displayed on the right, determine:

1. The DP characteristic of the composite two-terminal with nodes A and B
2. The power delivered by the same two-terminal when connected to a current source with impressed current $I_S = 2A$

3.12 Find the voltage v in the circuit shown in Fig. 3.60a. Also find the numerical solution for $G_1 = 6 \text{ m}\Omega^{-1}$, $G_2 = 13 \text{ m}\Omega^{-1}$, $G_3 = 10 \text{ m}\Omega^{-1}$ ($R_i = 1/G_i$, $i = 1, 2, 3$), $E = 24$ V, $A = 1$ mA.

3.13 Find the Thévenin equivalent representation shown in Fig. 3.26a for the two-terminal shown in Fig. 3.60b. Also find the numerical solution for $R = 25\Omega$, $A = 100$ mA.

References

1. Maxwell JC (1954, first publ 1891) A treatise on electricity and magnetism, vol I, 3rd edn. Dover Publications, New York (1891)
2. Lorentz H (1896) The theorem of Poynting concerning the energy in the electromagnetic field and two general propositions concerning the propagation of light. Versl Kon Akad Wentensch Amst 4:176
3. Helmholtz H (1853) Über einige Gesetze der Vertheilung Elektrischer Ströme in Körperlichen Leitem, mit Anwendung auf die Thierisch-Elektrischen Versuche (Some laws concerning the distribution of electrical currents in conductors with applications to experiments on animal electricity). Ann Phys Chem 89:211–233
4. Thévenin L (1883) Extension de la loi d'Ohm aux circuits électromoteurs complexes (Extension of Ohm's law to complex electromotive circuits). Ann Télégraphiques 10:222–224
5. Thévenin L (1883) Sur un nouveau théorème d'électricité dynamique (On a new theorem of dynamic electricity). Comptes rendus hebdomadaires des séances de l'Académie des Sci 97:159–161
6. Mayer HF (1926) Über das Ersatzschema der Verstärkerröhre (On equivalent circuits for electronic amplifiers). Telegr und Fernsprech Technik 15:335–337
7. Norton EL (1926) Design of finite networks for uniform frequency characteristic. Technical report TM26-0-1860, Bell Laboratories

Chapter 4
Advanced Concepts

Abstract In this chapter, we define the potential functions for two-terminal and n-terminal resistive elements. Moreover, we introduce some fundamental results concerning the variational principles for nonlinear resistive circuits.

4.1 Potential Functions

The concept of *potential function* or *dissipative function* for resistive elements originated from an idea of Maxwell [1], who formulated a "minimum heat" theorem for circuits containing only sources and two-terminal linear resistors. Much later, the result obtained by Maxwell was generalized by Millar [2] to circuits also containing two-terminal nonlinear resistors. The cited theorems state that voltages and currents settle to values such that the power dissipated by the resistors is minimum, as well as the amount of electric power converted to heat by the circuit. More in general, these values correspond to the stationarity (i.e., points where the gradient is null) of a suitable scalar potential function (minimum heat or dissipated power) that generalizes the Rayleigh dissipation function introduced in classical dynamics [3]. A potential function represents a very succinct description of the circuit and, as such, can be an important tool for the analysis of a circuit and of its properties. For instance, the deduction of the values of voltages and currents by a stationarity principle for a scalar function may suggest (especially in the nonlinear case) more efficient computational techniques than the classic ones, based on the solution of a system of algebraic equations. Furthermore, the use of a scalar function to describe the properties of a circuit can allow relating it to other kinds of physical systems, at first sight far from circuit theory. As an example, natural analogies can be established between some basic optimization problems and proper resistive circuits. Some important applications of these concepts are illustrated in Volume 2.

© Springer International Publishing AG 2018   93
M. Parodi and M. Storace, *Linear and Nonlinear Circuits:*
Basic & Advanced Concepts, Lecture Notes in Electrical Engineering 441,
DOI 10.1007/978-3-319-61234-8_4

Fig. 4.1 Symbol of a
nonlinear resistor

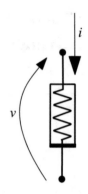

4.2 Content and Cocontent for Two-Terminal Memoryless Elements

Consider a memoryless two-terminal (henceforth simply called *resistor*), whose descriptive equation can be either linear or nonlinear. A nonlinear resistor is represented by the symbol of Fig. 4.1, with the descriptive variables v and i defined according to the standard choice.

4.2.1 Content Function

If the two-terminal resistor admits the current basis (i.e., it is current-controlled), its descriptive equation may be expressed in the form $v = g(i)$. (See Sect. 3.3.4.)

> The **content function** $G(i)$ of a current-controlled resistor is
>
> $$G(i) = \int_0^i g(x)\,dx \qquad (4.1)$$

The choice $i = 0$ for the lower limit of the integral is only customary; a change in the lower limit implies a new $G(i)$ that differs from the previous one only by a constant additive term. From definition Eq. 4.1, we obtain

$$v = \frac{dG}{di} \qquad (4.2)$$

whereby the lower limit in the integral Eq. 4.1 does not affect the determination of the voltage v. The relationship between the characteristic $v = g(i)$ and its content function $G(i)$ can be represented in the (i, v) plane as shown in Fig. 4.2.

Fig. 4.2 Relationship
between the DP
characteristic $v = g(i)$ and
the value of the content
function at $i = \hat{i}$

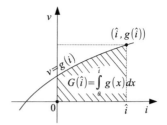

Notice that there are no restrictions for the energetic behavior of the considered resistors. For instance, the DP characteristic shown in Fig. 4.2 corresponds to an active two-terminal. Among the elements with content function, therefore, we can also include the ideal voltage source $v(t) = e(t)$, which admits the current basis and can be viewed as a time-varying resistive element for the algebraic nature of its descriptive equation.

Case Study 1: Linear Resistor

For a linear resistor R (Fig. 3.2) Ohm's law ($v = Ri = g(i)$) gives the content function $G(i)$ the following expression,

$$G(i) = \int_0^i Rx\,dx = \frac{1}{2}Ri^2 \qquad (4.3)$$

Case Study 2: Nonlinear Resistor

The descriptive equation for the nonlinear resistor whose DP characteristic is shown in Fig. 4.3 is:

$$v = \begin{cases} R_1 i & \text{for } i < 0 \quad \text{(branch I)} \\ R_2 i & \text{for } i \geq 0 \quad \text{(branch II)} \end{cases}$$

where R_1 and R_2 are the slopes of the branches I and II, respectively. The corresponding content function is

$$G(i) = \int_0^i g(x)\,dx = \begin{cases} \frac{1}{2}R_1 i^2 & \text{for } i < 0 \\ \frac{1}{2}R_2 i^2 & \text{for } i \geq 0 \end{cases}$$

Fig. 4.3 Case Study 2

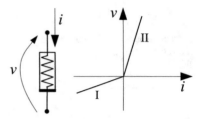

Case Study 3: Ideal Voltage Source

For the ideal voltage source $e(t)$ (Fig. 3.4a) the content function is time-varying, as a direct consequence of the time-dependence of $e(t)$:

$$G(i, t) = \int_0^i e(t)dx = e(t)i$$

Here the differential relationship between v and G must be expressed in terms of a *partial* derivative:

$$v = \frac{\partial G}{\partial i} = e(t).$$

4.2.2 Cocontent Function

If the two-terminal resistor admits the voltage basis (i.e., it is voltage-controlled), its descriptive equation can be formulated in the explicit form $i = f(v)$. (See Sect. 3.3.4.)

The cocontent function $\overline{G}(v)$ of a voltage-controlled resistor is

$$\overline{G}(v) = \int_0^v f(x)dx \qquad (4.4)$$

from which we immediately have

$$i = \frac{d\overline{G}}{dv}. \qquad (4.5)$$

For the lower limit of the integral Eq. 4.4, considerations completely analogous to those valid for the content function apply. The relationship between the characteristic $i = f(v)$ and its cocontent function $\overline{G}(v)$ is shown in Fig. 4.4. Among the elements

Fig. 4.4 Relationship
between the characteristic
$i = f(v)$ and the value of the
cocontent function at $v = \hat{v}$

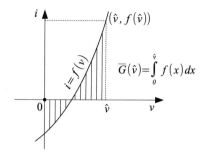

$$\overline{G}(\hat{v}) = \int_0^{\hat{v}} f(x)\,dx$$

with a cocontent function, we can also include the ideal current source $i(t) = a(t)$, which can be viewed as a (time-varying) resistive element and admits the voltage basis.

Case Study 1: Linear Resistor

Making reference to the voltage basis, we have $i = \frac{v}{R}$ and

$$\overline{G} = \int_0^v \frac{x}{R}dx = \frac{v^2}{2R} \tag{4.6}$$

Case Study 2: Nonlinear Resistor

The descriptive equation for the nonlinear resistor whose DP characteristic is shown in Fig. 4.5 is:

$$i = f(v) = \begin{cases} 0 & \text{for } v < 0 \quad \text{branch I} \\ \frac{v}{R} & \text{for } v \geq 0 \quad \text{branch II} \end{cases} \tag{4.7}$$

The corresponding cocontent function is:

$$\overline{G}(v) = \int_0^v f(x)dx = \begin{cases} 0 & \text{for } v < 0 \\ \frac{1}{2}\frac{v^2}{R} & \text{for } v \geq 0 \end{cases}$$

Fig. 4.5 Case Study 2

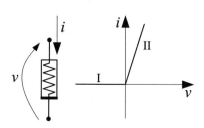

Case Study 3: Ideal Current Source

For the ideal current source $a(t)$ (Fig. 3.5a) the cocontent function is time-varying, as a direct consequence of the time-dependence of $a(t)$:

$$\overline{G}(v, t) = \int_0^v a(t)dx = a(t)v$$

Then the current i is obtained from $\overline{G}$ through a *partial* derivative:

$$i = \frac{\partial \overline{G}}{\partial v} = a(t).$$

4.2.3   Legendre Transformation

When a two-terminal resistive element admits both the voltage and the current basis, we can establish an important relationship between the resulting G and $\overline{G}$ functions. For this purpose, we represent the DP characteristic on the plane (i, v) and we choose a point $P_0 = (i_0, v_0)$ on this curve. Taking i_0 and v_0 as the lower limits for content and cocontent integrals, respectively, the values assumed by these integrals at a generic point $(\hat{i}, \hat{v})$ of the DP characteristic can be expressed as follows.

$$G(\hat{i}) = \int_{i_0}^{\hat{i}} g(i)di = \int_{i_0}^{\hat{i}} vdi \tag{4.8}$$

$$\overline{G}(\hat{v}) = \int_{v_0}^{\hat{v}} f(v)dv = \int_{v_0}^{\hat{v}} idv \tag{4.9}$$

where f is the inverse function of g. Then we have:

$$G(\hat{i}) + \overline{G}(\hat{v}) = \int_{i_0}^{\hat{i}} vdi + \int_{v_0}^{\hat{v}} idv = \int_{i_0 v_0}^{\hat{i}\hat{v}} d(iv) = \hat{i}\hat{v} - i_0 v_0$$

The terms $\hat{i}\hat{v}$ and $i_0 v_0$ denote the areas of the two rectangles shown in Fig. 4.6a. When P_0 is chosen at the intersection of the DP characteristic with either axis, as usually done (see the comments about the choice of the lower limits in the definitions of G and $\overline{G}$), we obviously have $i_0 v_0 = 0$ and

$$G(\hat{i}) + \overline{G}(\hat{v}) = \hat{i}\hat{v}$$

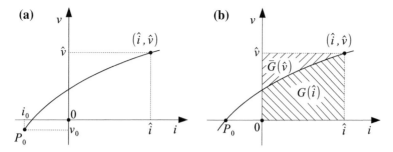

Fig. 4.6 a Geometrical elements for the Legendre transformation. **b** Geometrical relationship between G and $\overline{G}$ when $i_0 v_0 = 0$

which is the typical formulation for the Legendre transformation [4, 5]. This means that, at any instant, the sum of content and cocontent for a resistor equals the total power absorbed by that resistor in that instant.

It is easy to verify that, in this case, $G(\hat{i})$ and $\overline{G}(\hat{v})$ are complementary areas inside the rectangle with edges $\hat{i}$ and $\hat{v}$, as shown in Fig. 4.6b.

4.3   Content and Cocontent for Multiterminal Resistive Elements

Consider an n-terminal memoryless element (Fig. 4.7) and let $v = \begin{pmatrix} v_1 \\ \vdots \\ v_{n-1} \end{pmatrix}$ and

$i = \begin{pmatrix} i_1 \\ \vdots \\ i_{n-1} \end{pmatrix}$ be the vectors of the $(n-1)$ descriptive voltages and currents taken according to the standard choice.

Fig. 4.7 Descriptive variables for an n-terminal resistor

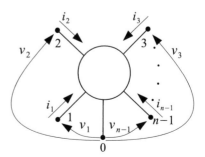

The n-terminal's descriptive equations in the current-controlled case can be written as $v = g(i)$, where g is a vector of $(n - 1)$ scalar functions. As a generalization of the result given for a two-terminal element,

> The **content function** for an n-terminal current-controlled resistor is defined by the integral
>
> $$G(i) = \int_0^i g^T(x)dx \qquad (4.10)$$

where the function to be integrated is the dot product between the vector of functions $g(x)$ and the vector of differential terms dx. The value $G(i)$ of the integral must be independent of the path connecting the points 0 and i; that is, the dot product $g^T(x)dx$ must be the exact differential dG of the content function. In order to fulfill this requirement, it is necessary and sufficient that, for any component g_k of the vector g and for any component i_j of the vector i, the following relation holds [6].

$$\frac{\partial g_k}{\partial i_j} = \frac{\partial g_j}{\partial i_k} \quad j, k = 1, \ldots, n - 1. \qquad (4.11)$$

These constraints on the descriptive equations are known as *conditions of local symmetry* [5].

In a completely analogous way,

> A voltage-controlled n-terminal element with descriptive equation $i = f(v)$ (i, v, f are vectors with $(n - 1)$ components) for which the conditions of local symmetry
>
> $$\frac{\partial f_k}{\partial v_j} = \frac{\partial f_j}{\partial v_k} \quad j, k = 1, \ldots, n - 1 \qquad (4.12)$$
>
> are fulfilled, admits the **cocontent function**
>
> $$\overline{G}(v) = \int_0^v f^T(x)dx. \qquad (4.13)$$

The cocontent function for the n-terminal element $\overline{G}(v)$ is independent of the integration path connecting the points 0 and v.

4.4 Additivity of Potential Functions

In this section we introduce a remarkable result [5] that settles the basis for the next sections.

✂ **Shortcut.** The additivity property introduced by the theorem is crucial, but the proof can be skipped without compromising the comprehension of the next sections.

Theorem 4.1 (Additivity of potential functions) *The potential function (content or cocontent) of a composite resistive element is given by the sum of the potential functions of its constituents.*

Proof The proof is given for the content case. Let us consider a circuit made up of current-controlled two-terminal resistive elements, including p voltage sources. The whole circuit can be schematically represented as in Fig. 4.8, where the voltage sources are brought outside a closed boundary, whereas all the remaining circuit elements are inside the boundary. The pair (v_k, i_k) denotes the descriptive variables of the kth resistor ($k = 1, \ldots, N$), defined according to the standard choice. On the contrary, the voltage sources are described by pairs $(e_j, \bar{i}_j)$ ($j = 1, \ldots, p$), chosen according to the nonstandard choice.

From Tellegen's theorem (Sect. 2.3) we know that the vector of voltages $v = (e_1, \ldots, e_p, v_1, \ldots, v_N)^T$ and the vector of currents $i = (-\bar{i}_1, \ldots, -\bar{i}_p, i_1, \ldots, i_N)^T$ are orthogonal. Then we have

$$\sum_{j=1}^{p} e_j \bar{i}_j = \sum_{k=1}^{N} v_k i_k. \tag{4.14}$$

This equation holds for the actual voltages and currents as well as for any set of voltages and currents compatible with the circuit graph. For the actual values, Eq. 4.14 has the obvious meaning of a circuit power balance: the l.h.s. represents the whole power delivered by the voltage sources and absorbed by the circuit portion inside the boundary, whereas the r.h.s. is the sum of powers absorbed by the resistive elements inside the boundary. Taking now a set of compatible currents,

$$-(\bar{i}_1 + d\bar{i}_1), \ldots, -(\bar{i}_p + d\bar{i}_p), i_1 + di_1, \ldots, i_N + di_N$$

Fig. 4.8 Circuit used to demonstrate the content additivity property

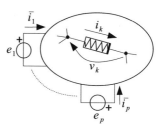

we can write

$$\sum_{j=1}^{p} e_j(\bar{i}_j + d\bar{i}_j) = \sum_{k=1}^{N} v_k(i_k + di_k). \tag{4.15}$$

Therefore, owing to Eq. 4.14, we obtain:

$$\sum_{j=1}^{p} e_j d\bar{i}_j = \sum_{k=1}^{N} v_k di_k. \tag{4.16}$$

Now we can integrate both members from 0 to i. The r.h.s. integral is the sum of the contents of the inner resistors. Then, the integral on the l.h.s can be viewed as the content G of the whole circuit portion inside the boundary. In other words, the content G of this $2p$-terminal is given by the sum of the contents of its elements.

As a final observation we recall that (Sect. 2.1.1) the descriptive variables for an $(m + 1)$-terminal component can be represented through the m branches forming any of its graphs. When such a component belongs to the circuit portion within the boundary of Fig. 4.8, we can formally replace it with the m branches of its graph, and we can associate a pair of descriptive variables (i_k, v_k) with each of these branches. After this replacement, we can follow the same procedure used to obtain Eq. 4.16. At the r.h.s. of this equation, m terms of the sum lead to the content function of the $(m + 1)$-terminal (current-controlled) component. Therefore, the content-additivity property can be directly extended to the case when parts of the resistive circuit constituents are current-based multiterminal components.   □

A completely analogous statement can be formulated and proved for the additivity of the cocontents [5]. These results are part of a set of general results concerning resistive circuits which are discussed in the next section.

4.5 Kirchhoff's Laws and Variational Principles for Potential Functions

In this section the relationships between Kirchhoff's laws and potential functions is pointed out by means of two *variational principles*.

> A **variational principle** is a method for determining the working condition of a physical system by identifying it as a stationary point (minimum, maximum, or saddle point) of a function.

> For a differentiable function of several real variables, a **stationary point** is an input (one value for each variable) where all its partial derivatives are zero (or, equivalently, the gradient is zero), that is, a point where the function "stops" increasing or decreasing (hence the name).

We start by introducing the so-called *stationary content principle* [7].

Theorem 4.2 (Stationary content principle) *For a resistive circuit admitting content function, the actual branch currents of the circuit make the circuit content stationary with respect to the graph-compatible current variations.*

Proof To prove this principle, we consider a circuit composed of m two-terminal elements (this restriction can be removed by following a line of reasoning similar to that of the preceding section). We assume that for each of them a content function can be given. Let v_k and i_k be the descriptive variables of the kth element, oriented according to the standard choice, and G_k its content function; the total content G of the circuit is

$$G = \sum_{k=1}^{m} G_k(i_k).$$

After partitioning the circuit graph into tree and cotree, the vectors v, i can be organized into the tree and cotree subvectors v_T, v_C, i_T, i_C. Taking into account the structure of the fundamental cut-set and loop matrices (Eqs. 2.1 and 2.5), we have

$$Ai = (\alpha \mid I) \begin{pmatrix} i_C \\ i_T \end{pmatrix} = \alpha i_C + i_T = 0; \quad Bv = (I \mid -\alpha^T) \begin{pmatrix} v_C \\ v_T \end{pmatrix} = v_C - \alpha^T v_T = 0$$

With this in mind, let us consider the first-order variation of G originated by graph-compatible variations δi_k (which are not required to be small) of the branch currents:

$$\delta G = \sum_{k=1}^{m} \frac{\partial G_k}{\partial i_k} \delta i_k = \sum_{k=1}^{m} v_k \delta i_k = \begin{pmatrix} v_T^T & v_C^T \end{pmatrix} \begin{pmatrix} \delta i_T \\ \delta i_C \end{pmatrix} = v_T^T \delta i_T + v_C^T \delta i_C. \quad (4.17)$$

Because the δi_C and δi_T variations must be compatible with the graph, we have $\alpha \delta i_C + \delta i_T = 0$. The substitutions $\delta i_T = -\alpha \delta i_C$ and $v_C^T = v_T^T \alpha$ into the r.h.s. of Eq. 4.17 give

$$\delta G = v_T^T (-\alpha \delta i_C) + v_T^T \alpha \delta i_C = 0.$$

This means that the change of G induced by graph-compatible variations of the branch currents is null; that is, G is stationary with respect to these variations. $\square$

As an example, consider the simple circuit shown in Fig. 4.9a. In terms of the currents i_E and i, the total content G of the circuit is $Ei_E + \frac{1}{2}Ri^2$. Following the

Fig. 4.9 Example concerning the stationarity content principle: **a** circuit; **b** content functions for the circuit components (*grey lines*) and circuit content function $G(i)$ (*black line*)

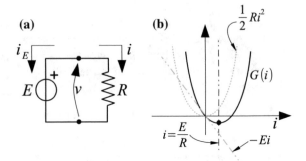

theorem proof, therefore we have $\delta G = \frac{\partial G}{\partial i_E}\delta i_E + \frac{\partial G}{\partial i}\delta i = E\delta i_E + Ri\delta i$. However, because $i_E = -i$, we have $\delta i_E = -\delta i$ and

$$\delta G = (-E + Ri)\delta i = 0 \quad \text{for} \quad i = \frac{E}{R}.$$

The condition $i = \frac{E}{R}$ follows directly from the KVL applied to the circuit. For this value of i, then, the content G is stationary with respect to a graph-compatible variation δi.

A geometrical interpretation of this result is possible by directly writing the content expression as $G(i) = -Ei + \frac{1}{2}Ri^2$. The function $G(i)$, which is plotted in Fig. 4.9b (black line), has its only minimum at the point $i = \frac{E}{R}$.

An analogous result, the *stationary cocontent principle*, holds for the circuit cocontent $\overline{G}$:

Theorem 4.3 (Stationary cocontent principle) *For a resistive circuit admitting a cocontent function, the actual branch voltages of the circuit make the circuit cocontent stationary with respect to the graph-compatible voltage variations.*

You can check your comprehension by proving this theorem.

The stationarity principles stated above are strictly related to Kirchhoff's laws. As shown in the following two case studies, the stationarity of the content function G implies that the KVLs on the circuit's fundamental loops are met, whereas the stationarity of the cocontent function $\overline{G}$ implies the KCLs' fulfillment on the circuit's fundamental cut-sets.

Fig. 4.10 Case Study 1

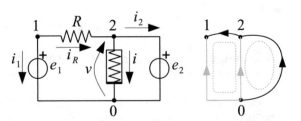

Case Study 1: Stationary Content Principle

We consider the circuit shown in Fig. 4.10 together with a possible tree (in grey) of its oriented graph. The two-terminal R is a linear resistor. The descriptive equation of the nonlinear resistor is $v = g(i)$. For each branch, the descriptive variables are taken according to the standard choice.

The content function of the circuit is:

$$G = e_1 i_1 + e_2 i_2 + \frac{1}{2} R i_R^2 + G_g(i) \quad \text{with } G_g(i) = \int_0^i g(x) dx$$

According to the chosen tree, the subvectors i_C, i_T are:

$$i_C = \begin{pmatrix} i_R \\ i_2 \end{pmatrix}; \quad i_T = \begin{pmatrix} i_1 \\ i \end{pmatrix} = \begin{pmatrix} -i_R \\ i_R - i_2 \end{pmatrix}$$

On this basis, we can write G as a function of the cotree currents only, that is:

$$G(i_R, i_2) = e_1(-i_R) + e_2 i_2 + \frac{1}{2} R i_R^2 + G_g(i_R - i_2).$$

The variation δG of the content function can be written as

$$\delta G = \frac{\partial G}{\partial i_R} \delta i_R + \frac{\partial G}{\partial i_2} \delta i_2 = (\nabla G)^T \begin{pmatrix} \delta i_R \\ \delta i_2 \end{pmatrix}$$

where the partial derivatives of G are given by:

$$\begin{cases} \dfrac{\partial G}{\partial i_R} = -e_1 + R i_R + \dfrac{\partial G_g}{\partial (i_R - i_2)} \cdot (+1) = \underbrace{-e_1 + R i_R + \overbrace{g(i_R - i_2)}^{v}}_{=0 \ (KVL)} \\[4mm] \dfrac{\partial G}{\partial i_2} = e_2 + \dfrac{\partial G_g}{\partial (i_R - i_2)} \cdot (-1) = \underbrace{e_2 - \overbrace{g(i_R - i_2)}^{v}}_{=0 \ (KVL)} \end{cases}$$

Both partial derivatives are null, due to the KVLs for the fundamental loops (the two inner loops). Then we have $\nabla G = \begin{pmatrix} 0 \\ 0 \end{pmatrix}$ and $\delta G = 0$, that is, a stationarity point of G, for any pair (i_R, i_2) that fulfills the KVL equations, as shown through the last equations.

As a concluding remark, we observe that the terms e_1, e_2, R in the expression of the content function $G(i_R, i_2)$ play the role of physical parameters. Any change of these parameters influences the shape of G and, obviously, the positions of the stationary points in the plane (i_R, i_2).

Fig. 4.11 Case Study 2

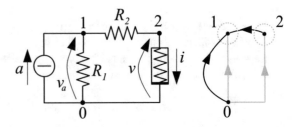

Case Study 2: Stationary Cocontent Principle

We consider the circuit shown in Fig. 4.11, where R_1 and R_2 are linear resistors. The grey lines in the circuit graph denote the tree. The nonlinear resistor is voltage-controlled, and its descriptive equation is $i = f(v)$. For each branch, the descriptive variables are taken according to the standard choice.

Taking as reference the voltages v_a and v of the tree, the cocontent function of the circuit can be directly written in the form:

$$\overline{G}(v_a, v) = -av_a + \frac{v_a^2}{2R_1} + \frac{(v_a - v)^2}{2R_2} + \overline{G}_f(v) \quad \text{with } \overline{G}_f(v) = \int_0^v f(x)\,dx$$

The variation $\delta\overline{G}$ of the cocontent function in terms of the variations δv_a, δv can be written as

$$\delta\overline{G} = \frac{\partial\overline{G}}{\partial v_a}\delta v_a + \frac{\partial\overline{G}}{\partial v}\delta v$$

and the stationarity condition $\delta\overline{G} = 0$ gives

$$\begin{cases} \frac{\partial\overline{G}}{\partial v_a} = -a + \frac{v_a}{R_1} + \frac{v_a - v}{R_2} = 0 \\ \frac{\partial\overline{G}}{\partial v} = -\frac{v_a - v}{R_2} + f(v) = 0 \end{cases} \tag{4.18}$$

which represent, as expected, the KCLs at nodes 1 and 2, respectively. Any pair (v_a, v) fulfilling Eq. 4.18 can be interpreted as a stationary point for the circuit.

Fig. 4.12 Example of
monotonic characteristic

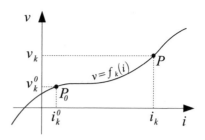

4.6   A Particular Variational Result: Resistive Circuits that Minimize Potential Functions

In this section we show that the content function of a circuit defined as in Sect. 4.4
(Fig. 4.8) and containing only resistors with monotonically increasing DP characteristics has its *minimum* for the actual current distribution.

Theorem 4.4 *In a circuit containing only independent sources and resistors with
monotonically increasing DP characteristics, the stationary point of the function G
for the actual current distribution is a minimum of G.*

✂ **Shortcut**. The theorem proof can be skipped without compromising the comprehension of the next section.

Proof We denote by $v_k = f_k(i_k)$ the descriptive equation ($k = 1, \ldots, N$) of the kth
resistor, with descriptive variables v_k and i_k. This means that, for any two points
P and P_0 with coordinates (i_k, v_k) and (i_k^0, v_k^0) on the characteristic (Fig. 4.12), the
inequality

$$(v_k - v_k^0)(i_k - i_k^0) \geq 0 \tag{4.19}$$

holds. Taking P_0 fixed and provisionally choosing for P the coordinates $i_k = i_k^0 +
di_k$; $v_k = f_k(i_k)$, the general inequality Eq. 4.19 becomes:

$$\left[f_k(i_k) - v_k^0\right] \underbrace{(i_k - i_k^0)}_{di_k} \geq 0.$$

We can now integrate with respect to the current, taking i_k^0 as the lower limit and a
generic i_k as the upper limit. By doing so, we obtain:

$$\int_{i_k^0}^{i_k} \left[f_k(i_k) - v_k^0\right] di_k \geq 0 \implies \underbrace{\int_{i_k^0}^{i_k} f_k(i_k) di_k}_{G(i_k) - G(i_k^0)} \geq \int_{i_k^0}^{i_k} v_k^0 di_k = v_k^0(i_k - i_k^0)$$

that is,

$$G(i_k) - G(i_k^0) \geq v_k^0(i_k - i_k^0).$$

Owing to the assumed resistors' characteristics, an inequality of this kind can be written for each of the N resistors within the boundary of Fig. 4.8. Each of them has its own reference point (i_k^0, v_k^0). By adding the N inequalities term by term, we find:

$$\sum_{k=1}^{N} G(i_k) - \sum_{k=1}^{N} G(i_k^0) \geq \sum_{k=1}^{N} v_k^0(i_k - i_k^0). \tag{4.20}$$

With the aim of applying Tellegen's theorem to the circuit, we first consider the actual voltages and currents in the circuit, say i_k^0, v_k^0 for the kth resistor $(k = 1, \ldots, N)$, without loss of generality. According to Fig. 4.8, the actual voltages and currents are:

$$e_1 \cdots e_p \ v_1^0 \cdots v_N^0$$

$$-i_1^0 \cdots -i_p^0 \ i_1^0 \cdots i_N^0.$$

Then we consider a second set of currents, not measured on the circuit but compatible with its graph (i.e., they obey KCLs):

$$-\overline{i_1} \cdots -\overline{i_p} \ i_1 \cdots i_N.$$

We can now apply Tellegen's theorem (Sect. 2.3) taking the set of actual voltages and the set of currents obtained by subtracting the current elements of the first set from those of the second:

$$\sum_{j=1}^{p} e_j(\overline{i_j} - \overline{i_j^0}) = \sum_{k=1}^{N} v_k^0(i_k - i_k^0). \tag{4.21}$$

From Eqs. 4.20 and 4.21, we obtain:

$$\sum_{k=1}^{N} G(i_k) - \sum_{k=1}^{N} G(i_k^0) \geq \sum_{j=1}^{p} e_j(\overline{i_j} - \overline{i_j^0})$$

which can be recast as

$$\sum_{k=1}^{N} G(i_k^0) - \sum_{j=1}^{p} e_j \overline{i_j^0} \leq \sum_{k=1}^{N} G(i_k) - \sum_{j=1}^{p} e_j \overline{i_j}. \tag{4.22}$$

This inequality shows that, under variations of the currents i_k, the r.h.s. is minimum for the actual current distribution [5].   □

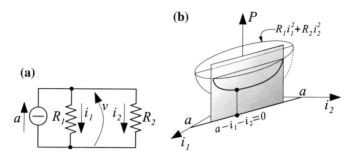

Fig. 4.13   a Circuit for the Maxwell theorem Case Study. **b** Geometric representation for the surface $P(i_1, i_2)$ and for the KCL constraint on i_1, i_2

This result includes, as a special case, the minimum heat theorem due to Maxwell, described in the next section.

4.7   Minimum Heat Theorem

In the special case when all resistors are linear, their content and cocontent functions coincide with the dissipated power and the previously discussed stationary principles become equivalent to the *minimum heat theorem* formulated by Maxwell.

Theorem 4.5 (Minimum heat theorem) *In a circuit composed of linear passive resistors and sources, the currents are such that the dissipated power is minimum.*

Although this result can now be considered as a special case of the stationary principles discussed in the previous sections, it is worth considering a simple case study to show how the problem of determining the currents in a circuit of this kind can be reformulated as a minimization problem.

Case Study
We consider the circuit shown in Fig. 4.13a, which is a resistive current divider, as discussed in Sect. 3.6.2. The power P absorbed (i.e., dissipated) by the linear resistors R_1 and R_2 is

$$P(i_1, i_2) = R_1 i_1^2 + R_2 i_2^2$$

At the same time, the resistor currents i_1 and i_2 are subject to a constraint imposed by the KCL, namely:

$$a - i_1 - i_2 = 0. \tag{4.23}$$

The minimum of P must be sought under this constraint. Indicating with λ a Lagrange multiplier, therefore, we consider the function:

$$F(i_1, i_2; \lambda) = P - \lambda(a - i_1 - i_2).$$

Both P and the KCL constraint have a simple geometric interpretation, shown in Fig. 4.13b. The minimum of P is the minimum of the curve intersection between the surface $P(i_1, i_2)$ and the vertical plane $a - i_1 - i_2 = 0$. Setting to zero the partial derivatives of F with respect to the variables i_1 and i_2 (stationary condition), we obtain

$$\begin{cases} \frac{\partial F}{\partial i_1} = 2R_1 i_1 + \lambda = 0 \\ \frac{\partial F}{\partial i_2} = 2R_2 i_2 + \lambda = 0 \end{cases} \Rightarrow \lambda = -2R_1 i_1 = -2R_2 i_2$$

which implies $R_1 i_1 = R_2 i_2$. Owing to both this result and the constraint Eq. 4.23, the minimum value of the power P occurs at the point with coordinates

$$i_1 = \frac{aR_2}{R_1 + R_2}; \quad i_2 = \frac{R_1}{R_2} i_1 = \frac{aR_1}{R_1 + R_2}.$$

Note that the products $R_1 i_1 = R_2 i_2$ are nothing but the voltage v on R_1 and R_2.

References

1. Maxwell JC (1954, first publ 1891) A treatise on electricity and magnetism, vol I, 3rd edn. Dover Publications, New York
2. Millar W (1951) Some general theorems for nonlinear systems possessing resistance. Philos Mag 42:1150–1160
3. Gantmacher F (1970) Lectures in analytical mechanics. MIR Publishers, Moscow
4. Stern TE (1965) Theory of nonlinear networks and systems: an introduction. Addison-Wesley, Boston
5. Penfield P, Spence R, Duinker S (1970) Tellegen's theorem and electrical networks. Research monograph, vol 58. The M.I.T. Press, Cambridge
6. Kreyszig E (1999) Advanced engineering mathematics, 8th edn. Wiley, New York
7. MacFarlane AGJ (1970) Dynamical system models. George G. Harrap, London

Part III
Memoryless Multi-ports: Descriptive Equations and Properties

Chapter 5
Basic Concepts

Live as if you were to die tomorrow. Learn as if you were to live forever.

Mahatma Gandhi

Abstract This chapter is focused on the concept of "port". Generic n-port components are introduced and linear, time-invariant, memoryless two-ports are defined and characterized through matrices, properties, and connections. The two-port components introduced in this chapter model real devices widely used in practical applications.

5.1 Port and n-Ports

As shown in Chap. 1, in a two-terminal element the KCL imposes a constraint on the currents flowing in the terminals. This is the basis to define the concept of *port*.

> **Port**: A pair of accessible terminals of a circuit that meet the so-called *port condition*; that is, they behave as for a two-terminal.

A black-box composite two-terminal, containing various components properly connected, is a one-port element (Fig. 5.1a). For instance, an appliance is a one-port element whose port is the plug.

Many common electronic devices and circuit blocks, such as transformers, electronic filters, and amplifiers have terminals that are naturally associated in pairs, each meeting the port condition. A component with $2n$ terminals associated in pairs is called an n-port and each port (owing to the port condition) needs as descriptive

© Springer International Publishing AG 2018

M. Parodi and M. Storace, *Linear and Nonlinear Circuits:
Basic & Advanced Concepts*, Lecture Notes in Electrical Engineering 441,
DOI 10.1007/978-3-319-61234-8_5

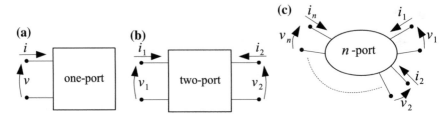

Fig. 5.1 One-port (**a**), two-port (**b**), and n-port (**c**) with standard choice of descriptive variables

variables one voltage and one current. Then, an n-port needs n port voltages and n port currents to be completely characterized. This also implies that the number of descriptive equations necessary to characterize an n-port is n. The standard choice for each port is the same as for a two-terminal, as shown in Fig. 5.1c. We notice that, if the component were viewed as a $2n$-terminal, it would require $(2n-1)$ descriptive voltages and as many descriptive currents. The restriction imposed by the concept of port introduces n independent constraints on the $2n$ terminal currents and allows neglecting the voltages corresponding to pairs of terminals belonging to different ports because they can remain undetermined without any consequence on the component's electrical behavior.

As shown in Sect. 3.4, a specific class of composite two-terminals/one-ports might admit a Thévenin or Norton equivalent representation. As for a composite two-terminal, we can define macromodels describing an n-port with descriptive equations expressed in terms of the port variables, as we show in Chap. 8. This allows us to reduce the complexity of circuit analysis, inasmuch as the n-port is regarded once more as a black- box connected to the outside world through its ports, which are often points where input signals are applied, output signals taken, or just contact points allowing the exchange of electrical energy. The port variables are the only variables to be considered in determining the n-port's response to applied signals. Of course, the knowledge of the neglected voltages is not necessary to determine the n-port's behavior.

In terms of standard-choice descriptive variables (Fig. 5.1c), the **power absorbed by an n-port** is the sum of the powers absorbed by each port; that is:

$$p(t) = \sum_{k=1}^{n} v_k i_k \tag{5.1}$$

Note for the readers who are aware of graphs (Sect. 2.1): The graph of an n-port is disconnected and is made up of the $2n$ nodes of the n-port and the n directed edges corresponding to the port voltages.

A memoryless *n*-port can admit the voltage basis (voltage-controlled *n*-port), the current basis (current-controlled *n*-port), or mixed bases, containing in any case one variable per port, because it is not possible to impose both voltage and current on the same port.

Two-ports (an example is shown in Fig. 5.1b) are the most widely used multiport elements. When both port voltages are referred to the same node (i.e., the lower terminals are connected by a short circuit), the two-port can be represented as a three-terminal. The possible bases for memoryless two-ports are: (i_1, i_2), (v_1, v_2), (v_1, i_2), (i_1, v_2).

5.2 Descriptive Equations of Some Two-Ports

Here we introduce the descriptive equations for some two-ports largely used to model electronic components and in particular *amplifiers*. An amplifier (from the Latin *amplificare*, "to enlarge or expand") is an electronic device that can increase the power of a signal by taking energy from a power supply and controlling the output to match the input signal shape but with a larger amplitude. In the next subsections we define a class of two-ports called *controlled sources* and the so-called *nullor*.

5.2.1 Controlled Sources

A *controlled source* (or dependent source) is a voltage or current source (port 2) whose value depends on a voltage or current somewhere else in the circuit (port 1). Port 1 is also called the *driving port* and port 2 the *driven port*.

Controlled sources are models of commonly used electronic components, simple or composite. Practical circuit elements have properties such as finite power capacity, voltage, current, or frequency limits that mean a model is valid only under specific assumptions.

The four possible types of *linear* controlled sources are defined in the following sections. All of them are time-invariant, memoryless, nonreciprocal, and active components. Dependent sources are not necessarily linear, as shown in Sect. 5.2.1.5.

5.2.1.1 Current-Controlled Current Source (CCCS)

The current-controlled current source (CCCS) is a two-port (shown in Fig. 5.2) often used to model current amplifiers and BJTs.[1]

---

[1]In the BJT case, the two ports share the same reference as for the port voltages, according to the three-terminal nature of this component.

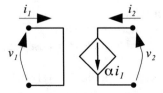

Fig. 5.2 Current-controlled current source (CCCS)

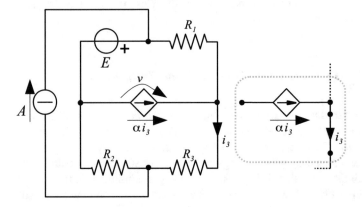

Fig. 5.3 Case Study for CCCS (*left circuit*) and detail (within the *dotted curve* on the *right*) showing the driving port of the CCCS

The **CCCS descriptive equations** are:

$$v_1(t) = 0$$
$$i_2(t) = \alpha i_1(t)$$
(5.2)

α is a dimensionless parameter called the *gain*, which can be positive or negative (if $\alpha = 0$ the port 2 of the CCCS degenerates to an open circuit), and $i_1(t)$ is called the *driving current*. Usually, $|\alpha| > 1$, at least when modeling amplifiers.

From the descriptive equations, it is apparent that the only admitted basis is the mixed basis (i_1, v_2).

The absorbed power is $p(t) = v_1 i_1 + v_2 i_2 = v_2 i_2 = \alpha i_1 v_2$. Because we cannot say anything a priori about the sign of this power, this two-port is active. It can easily be checked that it is nonreciprocal.

Remark: The port 1 of the CCCS (as well as of the other current-controlled source, the CCVS) is usually "hidden" in the circuit. It is just the terminal through which the driving current flows, as shown in Fig. 5.3. (See the detail on the right side.)

(a) 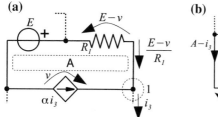   **(b)**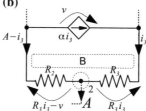

Fig. 5.4 Solution of the Case Study for CCCS

Fig. 5.5 Current-controlled
voltage source (CCVS)

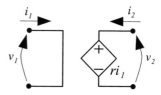

Case Study

Find the symbolic expression and the numerical value of the voltage v in the circuit shown in Fig. 5.3, with $E = 2$ V, $A = 2$ mA, $R_1 = 1$ kΩ, $R_2 = 3$ kΩ, $R_3 = 6$ kΩ, $\alpha = 2$.

As usual, we start by finding the variables that can be expressed in terms of the given unknowns, that is, v (problem target) and i_3 (driving variable): the voltage on R_1 (Fig. 5.4a) is $E - v$ (KVL for mesh A) and the voltage on R_3 (taken according to the standard choice) is $R_3 i_3$ (Ohm's law). Then, we obtain the current in R_1 (Ohm's law) and, by using the KCL for cut-set 1, we find $v = E + R_1(\alpha - 1)i_3$.

By using KVL for mesh B, Ohm's law, and KCL for cut-set 2 (Fig. 5.4b), we find $v = (R_2 + R_3)i_3 - R_2 A$. Finally, by solving the two equations, we find $v = E + R_1(\alpha - 1)\dfrac{E + R_2 A}{R_1(1 - \alpha) + R_2 + R_3} = 3$V.

5.2.1.2   Current-Controlled Voltage Source (CCVS)

The current-controlled voltage source (CCVS) is the two-port shown in Fig. 5.5.

Fig. 5.6 Case Study for
CCVS

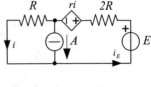

Fig. 5.7 Solution of the
Case Study for CCVS

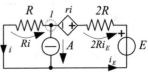

The **CCVS descriptive equations** are:

$$v_1(t) = 0$$
$$v_2(t) = ri_1(t)$$

(5.3)

r is the CCVS *gain* or *transresistance* (with physical dimension of Ω), which can be positive or negative (if $r = 0$ the port 2 of the CCVS degenerates to a short circuit), and $i_1(t)$ is the *driving current*.

From the descriptive equations, it is apparent that the only admitted basis is the current basis (i_1, i_2). Moreover, you can easily check that this two-port is also active and nonreciprocal.

Case Study

Find the symbolic expression and the numerical value of the current i_E in the circuit shown in Fig. 5.6, with $E = 5$ V, $A = 5$ mA, $R = 200\Omega$, $r = 400\Omega$. Also find the numerical value of the power absorbed by the current source.

The problem unknowns are i_E (problem target) and i (driving variable): the voltage on $2R$ (Fig. 5.7) is $2Ri_E$ (Ohm's law) and the voltage on R (taken according to the standard choice) is Ri (Ohm's law). Then, by using the KCL for cut-set 1, we find $i_E = A + i$ and, through KVL for the outer loop, we obtain $(R + r)i + 2Ri_E = E$.

By solving these two equations, we find $i_E = \dfrac{E + (R + r)A}{3R + r} = 8$ mA.

The power absorbed by the current source is $p = A \cdot Ri = RA(i_E - A) = 3$ mW.

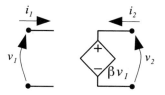

Fig. 5.8 Voltage-controlled voltage source (VCVS)

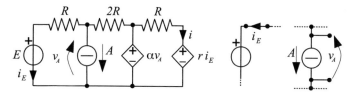

Fig. 5.9 Case Study for VCVS. On the right, the two driving variables and corresponding "hidden" ports are shown

5.2.1.3 Voltage-Controlled Voltage Source (VCVS)

The voltage-controlled voltage source (VCVS) is a two-port (shown in Fig. 5.8) used to model a voltage amplifier.

> The **VCVS descriptive equations** are:
>
> $$i_1(t) = 0$$
> $$v_2(t) = \beta v_1(t)$$
> (5.4)

β is the VCVS dimensionless *gain*, which can be positive or negative (if $\beta = 0$ the port 2 of the VCVS degenerates to a short circuit) and $v_1(t)$ is the *driving voltage*. Usually, $|\beta| > 1$, at least when modeling amplifiers.

From the descriptive equations, it is apparent that the only admitted basis is the mixed basis (v_1, i_2). Moreover, you can easily check that this two-port is also active and nonreciprocal.

Remark: Port 1 of the VCVS (as well as of the other voltage-controlled source, the VCCS) is just the pair of nodes across which the driving voltage is taken, as shown on the right side of Fig. 5.9.

> **Case Study**
>
> Find the symbolic expression and the numerical value of the current i in the circuit shown in Fig. 5.9, with $\alpha = 2$, $E = 2$ V, $A = 10$ mA, $R = 100\Omega$, $r = 300\Omega$.

Fig. 5.10 Solution of Case
Study for VCVS

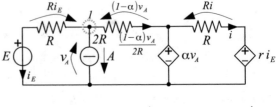

Fig. 5.11 Voltage-controlled
current source (VCCS)

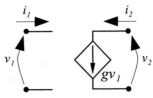

As usual, we start by finding the variables that can be expressed in terms
of the given unknowns, that is, i (problem target) and v_A and i_E (driving
variables), as shown in Fig. 5.10. Then, we obtain these three equations in the
three unknowns:

$$\alpha v_A = Ri + r i_E \quad \text{(KVL for the right inner loop)}$$
$$E + Ri_E = v_A \quad \text{(KVL for the left inner loop)}$$
$$i_E + A + \frac{1-\alpha}{2R}v_A = 0 \quad \text{(KCL for the nodal cut-set 1)}$$

By solving these three equations, we find

$$i = \frac{[2\alpha R + (1-\alpha)r]E + 2R(r-\alpha R)A}{(3-\alpha)R^2} = 40 \text{ mA}.$$

5.2.1.4   Voltage-Controlled Current Source (VCCS)

The voltage-controlled current source (VCCS) is the two-port shown in Fig. 5.11.

The **VCCS descriptive equations** are:

$$i_1(t) = 0$$
$$i_2(t) = gv_1(t) \tag{5.5}$$

g is the VCCS *gain* or *transconductance* (with the physical dimension of Ω^{-1})
and $v_1(t)$ is the *driving voltage*.

Fig. 5.12 Nonlinear CCVS

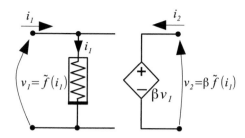

From the descriptive equations, it is apparent that the only admitted basis is the voltage basis (v_1, v_2). Moreover, you can easily check that this two-port is also active and nonreciprocal.

5.2.1.5 Nonlinear Controlled Sources

As stated above, dependent sources are not necessarily linear. However, a nonlinear controlled source can be easily obtained by properly combining a linear controlled source and a nonlinear resistor. For instance, a nonlinear CCVS with an equation of the driven port $v_2 = f(i_1)$, with f nonlinear function, can be obtained as shown in Fig. 5.12, by combining a linear VCVS and a nonlinear resistor admitting the current basis and described by the equation $v = \tilde{f}(i)$ (where $\tilde{f}(\cdot) = \frac{f(\cdot)}{\beta}$).

Indeed, $v_2 = \beta v_1 = \beta \tilde{f}(i_1) = f(i_1)$.

Similar considerations hold, *mutatis mutandis*, for any other nonlinear controlled source.

5.2.2 Nullor

The *nullor* is the two-port shown in Fig. 5.13.

Fig. 5.13 Nullor

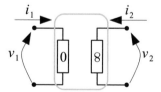

Fig. 5.14 Introductory circuit to understand the nullor's equations

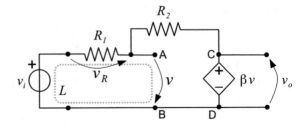

These apparently strange equations – that assign both variables at port 1, also called the *input port*, and leave completely free both variables at port 2, also called the *output port* – actually summarize the limit behavior of a linear VCVS inside a linear circuit.

In order to show this, we consider the introductory circuit shown in Fig. 5.14. The input port (terminals **A** and **B**) of the VCVS draws no current. For any finite value of the voltage gain parameter β, we easily obtain the expressions for the voltages v and v_o:

$$v = \frac{-v_i}{1 + (\beta + 1)\frac{R_1}{R_2}};$$

$$v_o = \beta v = \frac{-\beta}{1 + (\beta + 1)\frac{R_1}{R_2}} v_i \qquad (5.7)$$

For β values large enough to satisfy the inequality $(\beta + 1)\frac{R_1}{R_2} \gg 1$, the first of Eq. 5.7 gives $|v| \ll |v_i|$. This means that, in the KVL $v_i + v_R + v = 0$ (loop L), $|v|$ becomes negligible with respect to $|v_i|$ and $|v_R|$. Moreover, large values of β strongly simplify the v_o expression:

$$\lim_{\beta \to \infty} v = \lim_{\beta \to \infty} \frac{-v_i}{1 + (\beta + 1)\frac{R_1}{R_2}} = 0$$

$$\lim_{\beta \to \infty} v_o = \lim_{\beta \to \infty} \beta v = \frac{-\beta}{1 + (\beta + 1)\frac{R_1}{R_2}} v_i = -\frac{R_2}{R_1} v_i \qquad (5.8)$$

You can consider a realistic numerical case taking $\beta = 10^5$ and $\frac{R_1}{R_2} = 10^{-2}$. For these values, $v \simeq -10^{-3} v_i$ and $v_o \simeq -10^2 v_i = -\frac{R_2}{R_1} v_i$.

Summing up, for $\beta \to \infty$ the voltage and current at the input terminals **A**, **B** of the VCVS are zero. (Recall the nullor equations!) Moreover, v_o is proportional to

Fig. 5.15 Case Study 1

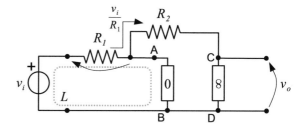

v_i through the coefficient $-\dfrac{R_2}{R_1}$, which depends only on components external to the VCVS.

This is just the circuit solution obtained by replacing the VCVS with a nullor, as in Case Study 1.

Case Study 1

Find the symbolic expressions of the voltage v_o in the circuit shown in Fig. 5.15.

We start by exploiting the nullor descriptive equations: the voltage on R_1 (Fig. 5.15) is v_i (KVL for mesh L) and the current in R_1 (Ohm's law) and R_2 (KCL for node 1) is v_i/R_1. Then, we obtain the voltage across R_2 (Ohm's law) and, by using the KVL for the outer loop, we find $v_o = -\frac{R_2}{R_1}v_i$.

In a linear circuit, then, the nullor models the VCVS behavior when its voltage gain becomes infinite. The nullor's symbol recalls the descriptive equations. On port 1 we have a degenerate two-terminal, called the *nullator*, described by Eq. 5.6. On port 2 we have another degenerate two-terminal, called the *norator*, with no descriptive equations. This means that the descriptive variables on the output port are determined by the rest of the circuit, because the nullor does not impose any constraint on them. Even if made up of two degenerate two-terminals, the nullor is not degenerate, because, in some sense, the anomalies of nullator and norator compensate each other. You can easily check that the nullor is a time-invariant, memoryless, active, nonreciprocal two-port.

The replacement of the VCVS with a nullor is commonly adopted as a good approximation for analyzing or designing electronic circuits containing *operational amplifiers*. An operational amplifier (op-amp) is shown in Fig. 5.16a, where $+V_S$ and $-V_S$ denote the positive and negative power supply, respectively. An op-amp is an amplifier circuit with very high ratio (on the order of hundreds of thousands of times) between the magnitude of output (v_2) and input (v_1) voltages. Though the term today commonly applies to integrated circuits, the original op-amp design used valves, and later designs used discrete transistor circuits. The labels "+" and "−" at the input

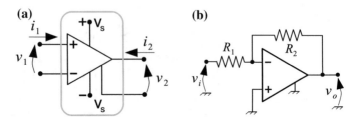

Fig. 5.16 Operational amplifier (**a**) and inverting amplifier (**b**). The symbol that appears four times in the lower part of the figure (**b**) denotes a reference node called the *ground*

Fig. 5.17 Case Study 2
(noninverting amplifier)

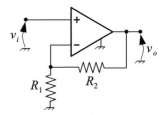

play an important role to ensure the *stability* of the overall circuit (a concept that is developed in Volume 2). The op-amp circuit corresponding to Fig. 5.15 is shown in Fig. 5.16b.

The apparently strange equations of the nullor actually summarize the two "golden rules" describing an ideal op-amp. The first rule only applies in the case (known as *closed-loop with negative feedback configuration*) where there is a signal path of some sort feeding back from the output to the inverting input, labeled by a "−" in Fig. 5.16b. This rule states that in this configuration the output attempts to do whatever is necessary to make the voltage v_1 zero. The second rule states that the input terminals draw no current (assumption of *infinite input impedance*). These rules are commonly used as a good first approximation for analyzing or designing op-amp circuits, thus the nullor (which implements these rules) is a commonly used model.

Case Study 2

    Find the symbolic expressions of the voltage v_o in the circuit shown in Fig. 5.17.

    We start once more by exploiting the nullor descriptive equations: the voltage on R_1 (Fig. 5.18) is v_i (KVL for mesh A) and the current in R_1 (Ohm's law) and R_2 (KCL for node 1) is v_i/R_1. Then, we obtain the voltage across R_2 (Ohm's law) and, by using the KVL for loop B, we find $v_o = \left(1 + \frac{R_2}{R_1}\right) v_i$.

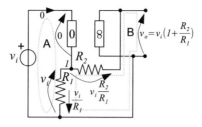

Fig. 5.18 Solution of Case Study 2

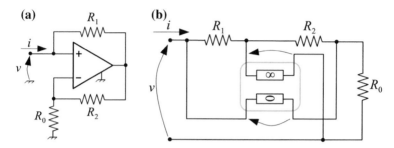

Fig. 5.19 Case Study 3 (negative resistance converter)

Case Study 3

Find the descriptive equation of the composite two-terminal shown in Fig. 5.19.

We start once more by exploiting the nullor descriptive equations: the current in R_1 (Fig. 5.20) is i (KCL for nodal cut-set 1) and the voltage across R_0 (KVL for loop A) is v. Moreover the voltage across R_2 (KVL for loop B) is $R_1 i$. Then, by using the KCL for nodal cut-set 2, we find $v = -\frac{R_0 R_1}{R_2} i$, which is the equation of a resistor with negative resistance, that is, a strictly active two-terminal. This possibility was anticipated in Sect. 3.2.1. This example shows a possible network realization of this apparently "strange" resistor using a composite two-terminal with an active component.

5.3 Matrix-Based Descriptions for Linear, Time-Invariant, and Memoryless Two-Ports

Under proper assumptions, linear, time-invariant, and memoryless two-ports can be compactly described by using some matrix-based descriptions. This specific class of two-ports is described by two homogeneous equations with real and constant

Fig. 5.20 Solution of Case
Study 3

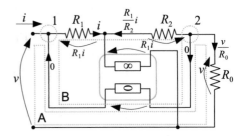

coefficients, with independent variables x_1 and x_2 and dependent variables y_1 and y_2:

$$\begin{pmatrix} y_1 \\ y_2 \end{pmatrix} = \begin{pmatrix} a & b \\ c & d \end{pmatrix} \begin{pmatrix} x_1 \\ x_2 \end{pmatrix} \tag{5.9}$$

In the following, the indices of the matrix elements refer, as usual, to rows and columns, not necessarily to the port number. Moreover, whenever there is no ambiguity, the matrices are simply denoted by a capital letter (e.g., R). Otherwise, the letter is inserted between brackets (e.g., (R) or $[R]$).

In Sect. 8.4 these descriptions are generalized to the nonhomogeneous (and then nonlinear) case. In Volume 2 a further generalization to circuits with memory is introduced.

5.3.1   Resistance Matrix

If the two-port admits the current basis, we can describe it through the *resistance matrix*:

$$\begin{pmatrix} v_1 \\ v_2 \end{pmatrix} = R \begin{pmatrix} i_1 \\ i_2 \end{pmatrix} = \begin{pmatrix} R_{11} & R_{12} \\ R_{21} & R_{22} \end{pmatrix} \begin{pmatrix} i_1 \\ i_2 \end{pmatrix} \tag{5.10}$$

Each matrix entry is a resistance that can be algebraically obtained as

$$R_{11} = \frac{v_1}{i_1}\bigg|_{i_2=0} \quad ; \quad R_{12} = \frac{v_1}{i_2}\bigg|_{i_1=0} \quad ; \quad R_{21} = \frac{v_2}{i_1}\bigg|_{i_2=0} \quad ; \quad R_{22} = \frac{v_2}{i_2}\bigg|_{i_1=0} \tag{5.11}$$

From a circuit perspective, this means that when port 2 of the composite two-port is left open ($i_2 = 0$; Fig. 5.21a), R_{11} can be obtained as the ratio between port-1 voltage v_1 (measured) and current i_1 (imposed), and R_{21} as the ratio between v_2 (measured) and i_1 (imposed). When port 1 is left open ($i_1 = 0$; Fig. 5.21b), we can similarly obtain R_{12} and R_{22}.

For instance, the resistance matrix for the CCVS is (from Eqs. 5.3)

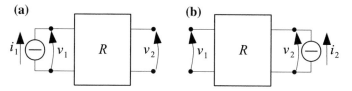

Fig. 5.21 Two-port with $i_2 = 0$ (**a**) and with $i_1 = 0$ (**b**)

Fig. 5.22 Case Study for resistance matrix

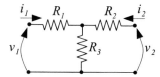

$$R = \begin{pmatrix} 0 & 0 \\ r & 0 \end{pmatrix}. \tag{5.12}$$

Case Study

Find the resistance matrix for the composite two-port shown in Fig. 5.22, with $R_1 = 100\Omega$, $R_2 = 30\Omega$, $R_3 = 50\Omega$.

Way 1. We find the descriptive equations of the two-port. By exploiting Kirchhoff's laws and Ohm's law (Fig. 5.23a), we obtain:

$$\begin{cases} v_1 = \underbrace{(R_1 + R_3)}_{R_{11}} i_1 + \underbrace{R_3}_{R_{12}} i_2 \\ v_2 = \underbrace{R_3}_{R_{21}} i_1 + \underbrace{(R_2 + R_3)}_{R_{22}} i_2 \end{cases} \tag{5.13}$$

Way 2. We find each matrix entry by using Eq. 5.11. The case with $i_2 = 0$ is shown in Fig. 5.23b, from which we find $v_1 = (R_1 + R_3)i_1$ and $v_2 = R_3 i_1$; that is, $R_{11} = R_1 + R_3 = 150\Omega$ and $R_{21} = R_3 = 50\Omega$. The case with $i_1 = 0$ is shown in Fig. 5.23c, from which we find $v_1 = R_3 i_2$ and $v_2 = (R_2 + R_3)i_2$; that is, $R_{12} = R_3 = 50\Omega$ and $R_{22} = R_2 + R_3 = 80\Omega$.

5.3.2 Conductance Matrix

If the two-port admits the voltage basis, we can describe it through the *conductance matrix*:

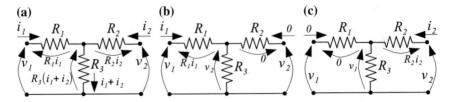

Fig. 5.23 Solution of Case Study. **a** Way 1. **b** Way 2, for $i_2 = 0$. **c** Way 2, for $i_1 = 0$

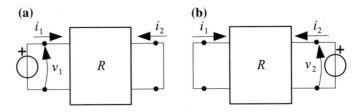

Fig. 5.24 Two-port with $v_2 = 0$ (**a**) and with $v_1 = 0$ (**b**)

$$\begin{pmatrix} i_1 \\ i_2 \end{pmatrix} = G \begin{pmatrix} v_1 \\ v_2 \end{pmatrix} = \begin{pmatrix} G_{11} & G_{12} \\ G_{21} & G_{22} \end{pmatrix} \begin{pmatrix} v_1 \\ v_2 \end{pmatrix} \tag{5.14}$$

Of course, if the two-port also admits the current basis, $R = G^{-1}$.

Each matrix entry is a conductance that can be algebraically obtained:

$$G_{11} = \left.\frac{i_1}{v_1}\right|_{v_2=0} ; \quad G_{12} = \left.\frac{i_1}{v_2}\right|_{v_1=0} ; \quad G_{21} = \left.\frac{i_2}{v_1}\right|_{v_2=0} ; \quad G_{22} = \left.\frac{i_2}{v_2}\right|_{v_1=0} \tag{5.15}$$

From a circuit perspective, this means that when port 2 of the composite two-port is short-circuited ($v_2 = 0$; Fig. 5.24a), G_{11} can be obtained as the ratio between port-1 current i_1 (measured) and voltage v_1 (imposed), and G_{21} as the ratio between i_2 (measured) and v_1 (imposed). When port 1 is short-circuited ($v_1 = 0$; Fig. 5.24b), we can similarly obtain G_{12} and G_{22}.

For instance, the conductance matrix for the VCCS is (from Eq. 5.5):

$$G = \begin{pmatrix} 0 & 0 \\ g & 0 \end{pmatrix}. \tag{5.16}$$

Case Study

    Find the conductance matrix (symbolic expression) for the composite two-port shown in Fig. 5.22.

Way 1. The descriptive Eq. 5.13 can be recast to express port currents in terms of port voltages. As an alternative, one can find these equations directly by once more using Kirchhoff's laws and Ohm's law. In both cases, we find:

$$
\begin{cases}
i_1 = \underbrace{\dfrac{(R_2 + R_3)}{R_{eq}^2}}_{G_{11}} v_1 - \underbrace{\dfrac{R_3}{R_{eq}^2}}_{G_{12}} v_2 \\[4mm]
i_2 = -\underbrace{\dfrac{R_3}{R_{eq}^2}}_{G_{21}} v_1 + \underbrace{\dfrac{(R_1 + R_3)}{R_{eq}^2}}_{G_{22}} v_2
\end{cases}
$$

where $R_{eq}^2 = R_1 R_2 + R_1 R_3 + R_2 R_3$.

Way 2. We find each matrix entry by using Eq. 5.11. In the case with $v_2 = 0$ we find $i_1 = \dfrac{(R_2 + R_3)}{R_{eq}^2} v_1$ and $i_2 = -\dfrac{R_3}{R_{eq}^2} v_1$, i.e., $G_{11} = \dfrac{(R_2 + R_3)}{R_{eq}^2}$ and $G_{21} = -\dfrac{R_3}{R_{eq}^2}$. With $v_1 = 0$ we find $i_1 = -\dfrac{R_3}{R_{eq}^2} v_2$ and $i_2 = \dfrac{(R_1 + R_3)}{R_{eq}^2} v_2$, i.e., $G_{12} = -\dfrac{R_3}{R_{eq}^2}$ and $G_{22} = \dfrac{(R_1 + R_3)}{R_{eq}^2}$. You can easily verify that $G = R^{-1}$.

5.3.3 Hybrid Matrices

If the two-port admits the mixed basis (i_1, v_2), we can describe it through the first hybrid matrix:

$$
\begin{pmatrix} v_1 \\ i_2 \end{pmatrix} = H \begin{pmatrix} i_1 \\ v_2 \end{pmatrix} = \begin{pmatrix} H_{11} & H_{12} \\ H_{21} & H_{22} \end{pmatrix} \begin{pmatrix} i_1 \\ v_2 \end{pmatrix} \tag{5.17}
$$

whose elements have hybrid physical dimensions (H_{11} has physical dimension of Ω, H_{22} of Ω^{-1}, whereas H_{12} and H_{21} are dimensionless) and can be algebraically obtained as

$$
H_{11} = \left.\frac{v_1}{i_1}\right|_{v_2=0} \; ; \quad H_{12} = \left.\frac{v_1}{v_2}\right|_{i_1=0} \; ; \quad H_{21} = \left.\frac{i_2}{i_1}\right|_{v_2=0} \; ; \quad H_{22} = \left.\frac{i_2}{v_2}\right|_{i_1=0} \tag{5.18}
$$

From a circuit perspective, the matrix entries can be computed by applying a method completely similar to the one described for matrices R and G, *mutatis mutandis*.

For instance, the H matrix for the CCCS is (from Eq. 5.2) $H = \begin{pmatrix} 0 & 0 \\ \alpha & 0 \end{pmatrix}$.

Completely analogous considerations hold for the second hybrid matrix, which exists if the two-port admits the other mixed basis (v_1, i_2):

$$\begin{pmatrix} i_1 \\ v_2 \end{pmatrix} = H' \begin{pmatrix} v_1 \\ i_2 \end{pmatrix} = \begin{pmatrix} H'_{11} & H'_{12} \\ H'_{21} & H'_{22} \end{pmatrix} \begin{pmatrix} v_1 \\ i_2 \end{pmatrix} \tag{5.19}$$

For instance, the H' matrix for the VCVS is (from Eq. 5.4) $H' = \begin{pmatrix} 0 & 0 \\ \beta & 0 \end{pmatrix}$.

Of course, if the two-port admits both the mixed bases, $H' = H^{-1}$.

You can check your comprehension by finding the hybrid matrices for the composite two-port shown in Fig. 5.22.

5.3.4 Transmission Matrices

Differently from the previous representations, each related to a basis, in the two representations described in this section the independent variables of Eq. 5.9 are taken both on the same port. If the "input" variables are those on the second port,[2] we can describe the two-port through the forward transmission matrix:

$$\begin{pmatrix} v_1 \\ i_1 \end{pmatrix} = T \begin{pmatrix} v_2 \\ -i_2 \end{pmatrix} = \begin{pmatrix} T_{11} & T_{12} \\ T_{21} & T_{22} \end{pmatrix} \begin{pmatrix} v_2 \\ -i_2 \end{pmatrix} \tag{5.20}$$

whose elements have hybrid physical dimensions (T_{11} and T_{22} are dimensionless, T_{12} has physical dimension of Ω, T_{21} of Ω^{-1}), and can be algebraically obtained as follows.

$$T_{11} = \frac{v_1}{v_2}\bigg|_{i_2=0} \quad ; \quad T_{12} = -\frac{v_1}{i_2}\bigg|_{v_2=0} \quad ; \quad T_{21} = \frac{i_1}{v_2}\bigg|_{i_2=0} \quad ; \quad T_{22} = -\frac{i_1}{i_2}\bigg|_{v_2=0} \tag{5.21}$$

For instance, the T matrix for the four controlled sources are: $T_{CCCS} = \begin{pmatrix} 0 & 0 \\ 0 & -\frac{1}{\alpha} \end{pmatrix}$,

$T_{CCVS} = \begin{pmatrix} 0 & 0 \\ \frac{1}{r} & 0 \end{pmatrix}$, $T_{VCVS} = \begin{pmatrix} \frac{1}{\beta} & 0 \\ 0 & 0 \end{pmatrix}$, $T_{VCCS} = \begin{pmatrix} 0 & -\frac{1}{g} \\ 0 & 0 \end{pmatrix}$.

From a circuit perspective, the matrix entries cannot be computed directly as for the other matrices, inasmuch as we cannot impose both voltage and current on port 2. But the inverse of each entry can be computed also in the circuit, provided that this is compatible with the descriptive equations (i.e., provided that KCL and KVL are not violated): with this *caveat*, for instance, $\dfrac{1}{T_{11}} = \dfrac{v_2}{v_1}\bigg|_{i_2=0}$ can be computed (Fig. 5.25a)

---

[2]We remark that in this case the term "input" is intended only to refer to their position in the mathematical expression because it is physically infeasible assigning both voltage and current on the same port. This is the reason for using the quotation marks around the word input.

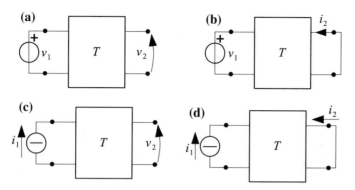

Fig. 5.25 Circuits used to compute the elements T_{11} (**a**), T_{12} (**b**), T_{21} (**c**), T_{22} (**d**)

by opening port 2 ($i_2 = 0$), imposing v_1 with a voltage source and measuring v_2. The other entries can be computed by following the same line of reasoning, as shown in Fig. 5.25. As a counterexample, in a current-controlled source (Figs. 5.2 and 5.5) we could not impose the voltage v_1 because the driving port is a short circuit, thus we could not compute T_{11} in this way.

Completely analogous considerations hold for the backward transmission matrix, which exists if the "input" variables are (v_1, i_1):

$$\begin{pmatrix} v_2 \\ -i_2 \end{pmatrix} = T' \begin{pmatrix} v_1 \\ i_1 \end{pmatrix} = \begin{pmatrix} T'_{11} & T'_{12} \\ T'_{21} & T'_{22} \end{pmatrix} \begin{pmatrix} v_1 \\ i_1 \end{pmatrix} \tag{5.22}$$

Of course, if the two-port admits both transmission matrices, $T' = T^{-1}$.

You can check your comprehension by finding the transmission matrices for the composite two-port shown in Fig. 5.22.

5.3.5 Generalizations to Higher Numbers of Ports

The representations defined in the previous subsections can be easily generalized for linear, time-invariant, memoryless n-ports with $n > 2$. In this case, each matrix size is $n \times n$.

Case Study

Determine if a three-port with resistance matrix $[R] = \begin{pmatrix} 0 & R & -R \\ -R & 0 & R \\ R & -R & 0 \end{pmatrix}$

is passive and if it admits the voltage basis.

The descriptive equations of the three-port are

$$\begin{cases} v_1 = Ri_2 - Ri_3 \\ v_2 = -Ri_1 + Ri_3 \\ v_3 = Ri_1 - Ri_2 \end{cases}$$

To evaluate its energetic behavior, we compute the absorbed power: $p = v_1i_1 + v_2i_2 + v_3i_3 = (Ri_2 - Ri_3)i_1 + (-Ri_1 + Ri_3)i_2 + (Ri_1 - Ri_2)i_3 = Ri_2i_1 - Ri_3i_1 - Ri_1i_2 + Ri_3i_2 + Ri_1i_3 - Ri_2i_3 = 0$. Then the three-port is nonenergic.

The voltage basis is admitted if and only if the conductance matrix is admitted, which means that the resistance matrix must be invertible. The determinant of matrix $[R]$ is $det([R]) = -R(-R^2) - RR^2 = 0$; thus matrix $[R]$ is not invertible and the voltage basis is not admitted.

5.4 Reciprocity of Two-Ports

The reciprocity of a two-port belonging to the class considered in the previous section can be simply assessed in terms of its matrix-based descriptions, if any.

5.4.1 Matrix R

In this case, denoting $v = \begin{pmatrix} v_1 \\ v_2 \end{pmatrix}$ and $i = \begin{pmatrix} i_1 \\ i_2 \end{pmatrix}$, we have

$$p' = (i')^T v'' = (i')^T Ri''$$

$$p'' = (v')^T i'' = (Ri')^T i'' = (i')^T R^T i''$$

(5.23)

Then, $p' = p''$ for any pair of electrical situations (Sect. 3.3.6) if and only if matrix R is symmetric, that is, if and only if $R_{12} = R_{21}$.

A completely similar condition ($G_{12} = G_{21}$) holds for matrix G (you can check it).

5.4.2 Matrix H

In this case, we have

$$p' = v_1'' i_1' + v_2'' i_2'$$

$$p'' = v_1' i_1'' + v_2' i_2''$$

and

$$p' = p'' \iff H_{11} i_1' i_1'' + H_{12} v_2'' i_1' + H_{21} v_2'' i_1' + H_{22} v_2' v_2'' =$$

$$= H_{11} i_1' i_1'' + H_{12} v_2' i_1'' + H_{21} v_2' i_1'' + H_{22} v_2' v_2''$$

$$\iff (H_{12} + H_{21}) v_2'' i_1' = (H_{12} + H_{21}) v_2' i_1''$$

Then, $p' = p''$ for any pair of electrical situations if and only if matrix H is skew-symmetric, that is, if and only if $H_{12} = -H_{21}$.

A completely similar condition $(H_{12}' = -H_{21}')$ holds for matrix H'. (You can check it.)

5.4.3 Matrix T

In this case, we have

$$p' = v_1'' i_1' + v_2'' i_2' = (T_{11} v_2'' - T_{12} i_2'')(T_{21} v_2' - T_{22} i_2') + v_2'' i_2'$$

$$p'' = v_1' i_1'' + v_2' i_2'' = (T_{11} v_2' - T_{12} i_2')(T_{21} v_2'' - T_{22} i_2'') + v_2' i_2''$$

and

$$p' - p'' = -T_{11} T_{22} v_2'' i_2' - T_{12} T_{21} v_2'' i_2' + v_2'' i_2' - (-T_{11} T_{22} v_2' i_2'' - T_{12} T_{21} i_2' v_2'' + v_2' i_2'') =$$

$$= (1 - T_{11} T_{22} + T_{12} T_{21})(v_2'' i_2' - v_2' i_2'') = 0$$

$$\iff T_{11} T_{22} - T_{12} T_{21} = 1$$

Then, $p' = p''$ if and only if the determinant of matrix T (denoted as $det(T)$) is 1.

A completely similar condition $(det(T') = 1)$ holds for matrix T'. (You can check it.)

Fig. 5.26 Case Study 1

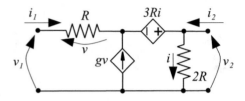

5.4.4   Remarks and Examples

We remark that to assess the reciprocity of a given two-port it is sufficient to check the reciprocity condition(s) for one of the admitted matrices. All the conditions are summarized in Sect. A.3 in the appendix.

The two-port analyzed in the Case Study of Sect. 5.3.1 is reciprocal, because $R_{12} = R_{21}$.

The controlled sources are not reciprocal, because in all cases $det(T) = 0$. For each controlled source, you can also check this property by looking at its other admitted matrix.

Case Study 1

Given the two-port shown in Fig. 5.26, find its matrix T and the value of the gain g (if any) ensuring reciprocity.

We can start by expressing the driving variables in terms of port variables: $i = \dfrac{v_2}{2R}$ and $v = Ri_1$. Then, we find $v_1 = Ri_1 - 3R\dfrac{v_2}{2R} + v_2$ (KVL for the outer loop) and $i_1 + gRi_1 + i_2 = \dfrac{v_2}{2R}$; that is, $i_1(1 + gR) = \dfrac{v_2}{2R} - i_2$. After a few algebraic manipulations, we obtain $i_1 = \dfrac{v_2}{2R(1 + gR)} - \dfrac{i_2}{1 + gR}$ and $v_1 = \dfrac{-gR}{2(1 + gR)}v_2 - \dfrac{R}{1 + gR}i_2$, corresponding to

$$T = \frac{1}{1 + gR}\begin{pmatrix} \dfrac{-gR}{2} & R \\ \dfrac{1}{2R} & 1 \end{pmatrix}$$

The reciprocity condition $det(T) = 1$ requires $g = -\dfrac{3}{R}$.

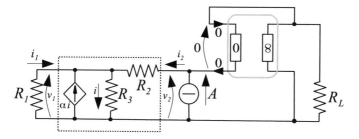

Fig. 5.27 Case Study 2

Case Study 2

   Given the circuit shown in Fig. 5.27, with $\alpha \neq 1$, find:

1. *The descriptive equations and (if admitted) the resistance matrix for the two-port within the dashed box*
2. *The condition on R_2 such that the same two-port does not admit the conductance matrix*
3. *For $\alpha = 0$, $R_1 = R_2 = R_3 = 1\,k\Omega$ and $A = \frac{2}{3}$ mA, the minimum value of R_L such that the power delivered by the nullor is lower than 1 W*

   You can easily check that the descriptive equations of the two-port within the dashed box are $v_1 = \dfrac{R_3}{(1-\alpha)}i_1 + \dfrac{R_3}{(1-\alpha)}i_2$ and $v_2 = \dfrac{R_3}{(1-\alpha)}i_1 + \left(\dfrac{R_3}{(1-\alpha)} + R_2\right)i_2$, corresponding to

$$R = \frac{R_3}{1-\alpha}\begin{pmatrix} 1 & 1 \\ 1 & 1 + \dfrac{R_2}{R_3}(1-\alpha) \end{pmatrix}$$

   For the second question, we have to impose that R is not invertible, that is, $det(R) = 0$, which implies $\dfrac{R_2}{R_3}(1-\alpha) = 0$. Inasmuch as $\alpha \neq 1$ this condition holds for $R_2 = 0$.

   For answering Question 3, we have to study the circuit shown in Fig. 5.28. From Ohm's law for the left resistor, we have $v_1 = -R_1i_1$ and from the two-port equations, we find $v_1 = R_1(i_1 + i_2) = R_1i_1 + R_1A$. Then, $i_1 = -A/2$ and $v_2 = R_1(i_1 + 2i_2) = 3R_1A/2$. The power delivered by the nullor is $p = -v_oi_o = \dfrac{v_o^2}{R_L}$. But $v_o = v_2 = 1$ V, then $p < 1$ W if and only if $R_L > 1\Omega$.

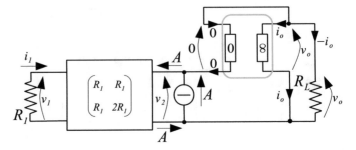

Fig. 5.28 Solution of Case Study 2, question 3

5.5 Symmetry of Two-Ports

We now define a further property that can characterize two-ports.

> A two-port is said to be **symmetrical** if its behavior does not change by swapping the two ports. Otherwise it is **nonsymmetrical**.

As for the reciprocity, the symmetry of a two-port belonging to the class considered in the previous section can also be simply assessed in terms of its matrix-based descriptions, if any.

All the conditions are summarized in Sect. A.3 in the appendix.

5.5.1 Matrix R

Given a two-port admitting matrix R, we exchange its ports and impose that its behavior does not change, according to the definition. The original two-port (shown in Fig. 5.29a) is described by Eq. 5.10.

The same two-port with exchanged ports (shown in Fig. 5.29b) is described by the descriptive equations:

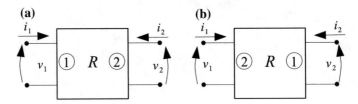

Fig. 5.29 Symmetry

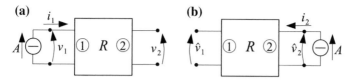

Fig. 5.30 Symmetry

$$\begin{pmatrix} v_1 \\ v_2 \end{pmatrix} = \begin{pmatrix} R_{22} & R_{21} \\ R_{12} & R_{11} \end{pmatrix} \begin{pmatrix} i_1 \\ i_2 \end{pmatrix}$$

Then, by comparing these descriptive equations with the original ones, we can conclude that the two-port is symmetrical if and only if $R_{12} = R_{21}$ and $R_{11} = R_{22}$. Notice that the symmetry of the two-port does not correspond simply to the symmetry of matrix R, which instead characterizes the reciprocity of the two-port.

For instance, if we take the circuit shown in Fig. 5.30a, we have $v_1 = R_{11}A$ and $v_2 = R_{21}A$. The same two-port connected as shown in Fig. 5.30b provides $\hat{v}_1 = R_{12}A$ and $\hat{v}_2 = R_{22}A$. The two-port is symmetrical if and only if $\hat{v}_1 = v_2$ and $\hat{v}_2 = v_1$, that is, if and only if the symmetry conditions hold.

Completely similar conditions ($G_{12} = G_{21}$ and $G_{11} = G_{22}$) hold for matrix G.

5.5.2 Hybrid Matrices

For a two-port admitting matrix H, we exchange its ports. Because $\begin{pmatrix} i_1 \\ v_2 \end{pmatrix} =$ $P \begin{pmatrix} v_2 \\ i_1 \end{pmatrix}$, where $P = \begin{pmatrix} 0 & 1 \\ 1 & 0 \end{pmatrix} = P^{-1}$, we have:

$$\begin{pmatrix} v_1 \\ i_2 \end{pmatrix} = H \begin{pmatrix} i_1 \\ v_2 \end{pmatrix} = HP \begin{pmatrix} v_2 \\ i_1 \end{pmatrix}$$

$$\begin{pmatrix} v_2 \\ i_1 \end{pmatrix} = (HP)^{-1} \begin{pmatrix} v_1 \\ i_2 \end{pmatrix}$$

The symmetry condition requires that $HP = (HP)^{-1} = P^{-1}H^{-1}$; that is, $H = P^{-1}H^{-1}P^{-1} = PH^{-1}P$. Then, because $H^{-1} = \dfrac{1}{det(H)} \begin{pmatrix} H_{22} & -H_{12} \\ -H_{21} & H_{11} \end{pmatrix}$ and $PH^{-1}P = \dfrac{1}{det(H)} \begin{pmatrix} H_{11} & -H_{21} \\ -H_{12} & H_{22} \end{pmatrix}$, the two-port is symmetrical if and only if

$$H_{11} = \frac{H_{11}}{det(H)}, \quad H_{12} = -\frac{H_{21}}{det(H)}, \quad H_{21} = -\frac{H_{12}}{det(H)}, \text{ and } H_{22} = \frac{H_{22}}{det(H)}.$$

These conditions are trivially satisfied for $det(H) = 1$ and $H_{21} = -H_{12}$. A less trivial solution is described in the case study below.

Completely similar conditions hold for matrix H'.

We remark that these symmetry conditions hold only if matrix H is invertible (i.e., $det(H) \neq 0$).

Case Study

Determine if and under which conditions a two-port described by a matrix H with $det(H) \neq 1$ and $H_{12} \neq -H_{21}$ can be symmetrical.

Because $det(H) \neq 1$ by assumption, the symmetry conditions $H_{11} = \dfrac{H_{11}}{det(H)}$ and $H_{22} = \dfrac{H_{22}}{det(H)}$ can be fulfilled only if $H_{11} = H_{22} = 0$. Then, $det(H) = -H_{12}H_{21}$ and the other two symmetry conditions become $H_{12} = -\dfrac{H_{21}}{-H_{12}H_{21}} = \dfrac{1}{H_{12}}$ and $H_{21} = -\dfrac{H_{12}}{-H_{12}H_{21}} = \dfrac{1}{H_{21}}$; that is, $H_{12}^2 = H_{21}^2 = 1$. Because $H_{12} \neq -H_{21}$ (i.e., the two-port is not reciprocal), the symmetry conditions are $H_{12} = H_{21} = 1$ or $H_{12} = H_{21} = -1$. In this case the two-port is symmetrical but not reciprocal.

5.5.3   Transmission Matrices

A two-port admitting matrix T is symmetrical if and only if $T_{11} = \dfrac{T_{22}}{det(T)}$, $T_{12} = \dfrac{T_{12}}{det(T)}$, $T_{21} = \dfrac{T_{21}}{det(T)}$ and $T_{22} = \dfrac{T_{11}}{det(T)}$. You can check your comprehension by deriving these conditions.

You can also check that there are only two possible matrix structures fulfilling these conditions: $T = \begin{pmatrix} T_{11} & T_{12} \\ T_{21} & -T_{11} \end{pmatrix}$, with $T_{11}^2 + T_{12}T_{21} = -1$ (in this case the two-port is also reciprocal, because $det(T) = 1$) and $T = \begin{pmatrix} \alpha & 0 \\ 0 & -\alpha \end{pmatrix}$, with $\alpha = 1$ or $\alpha = -1$.

Completely similar conditions hold for matrix T'.

We remark that these symmetry conditions hold only if matrix T is invertible (i.e., $det(T) \neq 0$).

5.6   Directionality of Two-Ports

Another property that can be used to characterize a two-port is the so-called *directionality*.

Fig. 5.31 Example of
zero-directional two-port,
with two uncoupled ports

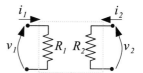

A two-port is **zero-directional** if the two ports are uncoupled. (See an example
in Fig. 5.31.) On the contrary, a two-port is **one-directional** if it is not zero-
directional and there exists one and only one relationship involving only the
descriptive variables of one port.

For instance, any pair of uncoupled one-port elements is a zero-directional two-
port, whereas the ideal controlled sources are one-directional two-ports.

Zero-directionality implies that the matrices associated with a basis are diagonal
and neither T nor T' are admitted.

One-directionality implies that in the matrices associated with a basis either the
element 12 (one-directionality $1 \rightarrow 2$) or the element 21 (one-directionality $2 \rightarrow 1$)
is zero. For example, the matrix R of Eq. 5.12 and the matrix G of Eq. 5.16 regard one-
directional $1 \rightarrow 2$ elements. It also implies that either $det(T) = 0$ or $det(T') = 0$,
depending on the versus of directionality.

5.7 Two-Port Ideal Power Transferitors

In this section, we deal with a particular subclass of nonenergic two-ports known as
ideal power transferitors. In generic nonenergic two-ports (Sect. 3.3.5 and Eq. 5.1),
the power absorbed by a single port may also be identically (i.e., for any electrical
situation) zero, whereas in the ideal power transferitors the overall power only is
identically zero.

An **ideal power transferitor** [1] is a nonenergic two-port that absorbs an
identically null overall instantaneous power $p(t) = v_1 i_1 + v_2 i_2$, inasmuch as
both $v_1 i_1$ and $v_2 i_2$ are not identically zero.

Of course, this constraint implies that $v_1/i_2 = -v_2/i_1$ or equivalently $v_1/v_2 =
-i_2/i_1$, but these equations are not sufficient to define a two-port (we would need a
pair of independent scalar equations).

We remark that a two-port ideal power transferitor always admits both T and T'.
Indeed, because $p_1 = v_1 i_1 \neq 0$ and $p_2 = v_2 i_2 \neq 0$, nonenergic zero-directional two-
ports whose ports are either open or short circuits cannot be ideal power transferitors.

Moreover, as $p_1 + p_2 = 0$, it is not possible to have $p_1 = 0$ and $p_2 \neq 0$ or vice versa. This implies that one-directional two-ports where one port is either an open or short circuit cannot be ideal power transferitors. For these reasons, *an ideal power transferitor cannot be either zero-directional or one-directional*. As a consequence (Sect. 5.6), T and T' are nonsingular.

In conclusion, without loss of generality, we choose matrix T to represent a generic ideal power transferitor. The overall instantaneous power $p(t)$ absorbed by such a two-port can be expressed through the entries of matrix T as a function of v_2 and i_2:

$$p = v_1 i_1 + v_2 i_2 = \underbrace{(T_{11}T_{21})}\ v_2^2 + \underbrace{(T_{12}T_{22})}\ i_2^2 + \underbrace{(1 - T_{11}T_{22} - T_{12}T_{21})}\ v_2 i_2 \quad (5.24)$$

The null-power condition requires that the three underbraced coefficients in Eq. 5.24 be null. Then, in order to find the matrices T that fulfill the null-power condition, we have to solve nonlinear system of three equations with respect to the four unknowns $T_{11}, T_{22}, T_{12}, T_{21}$:

$$\begin{cases} T_{11}T_{21} = 0 \\ T_{12}T_{22} = 0 \\ T_{11}T_{22} + T_{12}T_{21} = 1 \end{cases} \quad (5.25)$$

The first two of Eq. 5.25 constrain at least two of the unknowns (one between T_{11} and T_{21} and one between T_{12} and T_{22}) to vanish. There would be four cases, but the third equation excludes the possibility that both a diagonal and an off-diagonal entry may be zero; that is, only two of the four cases are admitted. In conclusion, it is only possible that either the two diagonal entries are both zero and the product of the two off-diagonal ones returns one or vice versa. Then, there are only two structurally different subsets of solutions of Eq. 5.25.

5.7.1 Ideal Transformer

The first solution subset is given by $T_{12} = T_{21} = 0$, $T_{11}T_{22} = 1$. This subset of solutions fulfills the reciprocity condition $det(T) = 1$ and corresponds to a two-port circuit element shown in Fig. 5.32a, called the *ideal transformer*, whose transmission matrix T is:

$$T = \begin{pmatrix} n & 0 \\ 0 & 1/n \end{pmatrix} \quad (5.26)$$

where n is a dimensionless real parameter called the *transformation ratio*.

Fig. 5.32 Ideal power
transferitors. **a** Ideal
transformer. **b** Gyrator

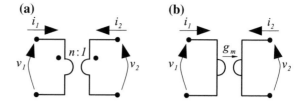

The **ideal transformer descriptive equations** are:

$$v_1(t) = nv_2(t)$$
$$i_2(t) = -ni_1(t)$$ (5.27)

Remark 1: The dots determine the orientation of each port; that is, they denote the upper terminal of each port.

Remark 2: The port at the "n" side is port 1.

From the descriptive equations, it is apparent that the only admitted bases are the two mixed bases (v_1, i_2) and (i_1, v_2). Moreover, this two-port is linear, time-invariant, memoryless, nonenergic, reciprocal, and nonsymmetrical (by assuming $n \neq 1$). If $n = 1$, we obtain the so-called *ideal connector*.

You can check your comprehension by finding one or more equivalent models of the ideal transformer, containing only controlled sources.

The two-port ideal transformer was proposed as an idealized model of a device (transformer) that has been well-known from more than a century, and is exploited to transfer electrical energy without significant losses. According to the historical notes on circuit theory reported in [2–4], the generalized N-port ideal transformer was introduced into the circuit literature in 1920 [5] as "an electrical device which may be used to transfer energy, but itself can neither absorb, store, nor supply energy" (Appendix VI in [5], where the basic concepts for the "new" element are established).

Among the many applications, real transformers are also used extensively in electronic devices to decrease the supply voltage to a level suitable for the low-voltage circuits they contain. In these cases, they appear as black-boxes placed between the plug and the device. Your daily experience makes it apparent that a real transformer differs from the ideal component for at least two reasons: it warms up when working – meaning that it is not an ideal power transferitor – and it does not work anymore when connected to a DC source – meaning that the model approximates the physical system reasonably well only under specific assumptions, for example, AC input within a proper frequency range.

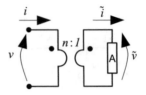

Fig. 5.33 Case Study 1 for ideal transformer

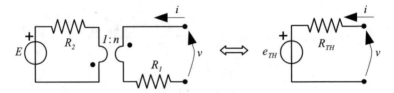

Fig. 5.34 Case Study 2 for ideal transformer

Case Study 1

Find the descriptive equation of the composite two-terminal shown in Fig. 5.33, where the two-terminal A connected to the ideal transformer is described by the implicit equation $f(\tilde{i}, \tilde{v}) = 0$.

From the ideal transformer equations, we obtain $v = n\tilde{v}$ and $i = \dfrac{\tilde{i}}{n}$ (notice that $\tilde{i}$ is the opposite of the port-2 current). Because $f(\tilde{i}, \tilde{v}) = 0$, we finally obtain that the descriptive equation of the composite two-terminal is $f(ni, \dfrac{v}{n}) = 0$. For instance, if the two-terminal A were a resistor, the descriptive equation of the composite two-terminal would be $v = n^2 R i$, that is, again a resistor, but with a scaled resistance value.

Case Study 2

Find the descriptive equation and the Thévenin equivalent of the composite two-terminal shown in Fig. 5.34.

First of all, we have to identify the ideal transformer ports (through the label n in the figure) and their orientations (through the dots) correctly. Then, as usual, we express the descriptive variables in terms of the problem variables, which in this case are v and i: thus (Fig. 5.35) we find the voltage across R_1 (Ohm's law), the current circulating in the left mesh (current equation of the ideal transformer), the voltage across R_2 (Ohm's law), and the voltage on port 2 of the ideal transformer (KVL). Next, by using the voltage equation of the

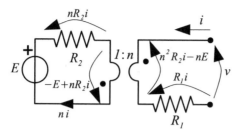

Fig. 5.35 Solution of Case Study 2

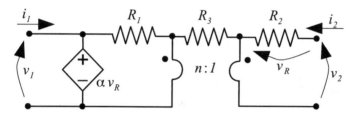

Fig. 5.36 Case Study 3 for ideal transformer

ideal transformer, we find the voltage on port 1 ($n^2 R_2 i - n E$) and, through KVL, the required descriptive equation: $v = R_1 i + n^2 R_2 i - n E$.

By simply reorganizing the descriptive equation, we can easily find the Thévenin equivalent: $v = \underbrace{(R_1 + n^2 R_2)}_{R_{TH}} i \underbrace{- n E}_{e_{TH}}$.

Case Study 3

For the composite two-port shown in Fig. 5.36, where $\alpha \neq n$, find the descriptive equations, the admitted bases and matrices, the resistance matrix (if admitted), the conditions on the parameters such that the composite two-port is reciprocal, symmetrical, passive, active, or nonenergic (check each property independently of the others).

As usual, we express the descriptive variables in terms of the problem variables, which in this case are the port variables and the driving voltage v_R: thus (Fig. 5.37) we find the voltage across R_2 (Ohm's law) and the voltage on port 2 of the ideal transformer (KVL). This implies that $v_R = -R_2 i_2$ and $v_1 = \alpha v_R = -\alpha R_2 i_2$, which is a first descriptive equation. Then, from cut-set 1, it is evident that the current flowing in R_3 must be null, as well as the voltage across it. This also implies that the current in port 2 of the ideal transformer is i_2. Next, by using the current equation of the ideal transformer, we find the current

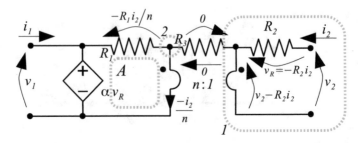

Fig. 5.37 Solution of Case Study 3

in its port 1 $(-i_2/n)$, which is also the current flowing in R_1 (due to KCL for nodal cut-set 2). Then the voltage across R_1 is $-R_1i_2/n$. The voltage equation of the ideal transformer states that the voltage on its port 1 is $n(v_2 - R_2i_2)$ and, through KVL for inner loop A, the second descriptive equation is: $-\alpha R_2i_2 = n(v_2 - R_2i_2) - R_1i_2/n$; that is, $v_2 = \left(\dfrac{n-\alpha}{n}R_2 + \dfrac{R_1}{n^2}\right)i_2$.

From the descriptive equations, the admitted bases are (i_1, i_2) (corresponding to matrix R) and (i_1, v_2) (corresponding to matrix H). Matrix T' is also admitted.

The resistance matrix can be directly identified from the descriptive equations:

$$R = \begin{pmatrix} 0 & -\alpha R_2 \\ 0 & \dfrac{n-\alpha}{n}R_2 + \dfrac{R_1}{n^2} \end{pmatrix}$$

Reciprocity: The two-port is reciprocal only for $\alpha = 0$ (such that $R_{12} = R_{21}$).

Symmetry: The symmetry conditions require $\alpha = 0$ and (to impose also $R_{11} = R_{22}$) $R_2 = -R_1/n^2$, which is not compatible with the usual assumption of positive resistances.

Energetic behavior: Owing to the descriptive equations, the absorbed power $(p(t) = v_1i_1 + v_2i_2)$ is in general a function of i_1 and another port variable, thus we cannot say anything about its sign. Then, for $\alpha \neq 0$, the two-port is active. For $\alpha = 0$, we have $p(t) = \left(R_2 + R_1/n^2\right)i_2^2$: in this case, the two-port is passive.

5.7.2   Gyrator

The second subset of solutions of Eq. 5.25 is given by $T_{11} = T_{22} = 0$, $T_{12} = T_{21} = 1$. This subset corresponds to a nonreciprocal two-port circuit element shown

Fig. 5.38 Case Study 1 for gyrator

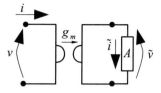

in Fig. 5.32b, called the *gyrator*. Its transmission matrix T is antidiagonal and such that $det(T) = -1$:

$$T = \begin{pmatrix} 0 & 1/g_m \\ g_m & 0 \end{pmatrix}$$ (5.28)

where g_m is a real parameter (with physical dimension of Ω^{-1}) called the *gyration conductance*.

The **gyrator descriptive equations** are:

$$\begin{aligned} i_1(t) &= g_m v_2(t) \\ i_2(t) &= -g_m v_1(t) \end{aligned}$$ (5.29)

From the descriptive equations, it is apparent that the only admitted bases are the voltage basis and the current basis. Moreover, this two-port is linear, time-invariant, memoryless, nonenergic, nonreciprocal, and nonsymmetrical. (You can easily check it.)

The (ideal) gyrator was conceived by Tellegen in 1948 [6], not as a model of a physical element but as the paradigm of nonreciprocal and passive two-port elements. It represents one of the most significant cases where circuit theory led to the definition of a new (ideal) element whose physical realization was obtained later on. (See, e.g., [7–9].) For this reason, we call the (ideal) gyrator simply the "gyrator," whereas the name "ideal transformer" refers to the introduction of this two-port as an idealization of an existing physical element. The gyrator basically changes the basis of a given two-terminal element, as shown in the following Case Study 1. The most interesting application of such a property was the possibility of converting an inductor into a capacitor (two components that are introduced in Volume 2) and vice versa.

Case Study 1

Find the descriptive equation of the composite two-terminal shown in Fig. 5.38, where the two-terminal A connected to the gyrator is described by the implicit equation $f(\tilde{i}, \tilde{v}) = 0$.

From the gyrator equations, we obtain $i = g_m \tilde{v}$ and $\tilde{i} = g_m v$. Because $f(\tilde{i}, \tilde{v}) = 0$, we finally obtain that the descriptive equation of the composite

Fig. 5.39 Case Study 2 for gyrator

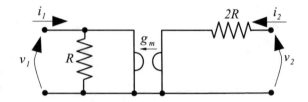

two-terminal is $f(g_m v, \dfrac{i}{g_m}) = 0$. For instance, if the two-terminal A were a linear resistor $(\tilde{v} = R\tilde{i})$, the descriptive equation of the composite two-terminal would be $i = g_m^2 Rv$, that is, again a resistor, but with a scaled resistance value. If the two-terminal A were a nonlinear resistor admitting the current basis only $(\tilde{v} = f(\tilde{i}))$, the descriptive equation of the composite two-terminal would be $i = g_m f(g_m v) = \hat{f}(v)$, that is, again a nonlinear resistor, but admitting the voltage basis only.

Case Study 2

For the composite two-port shown in Fig. 5.39, where $R > 0$, find descriptive equations, admitted bases and matrices, conductance matrix (if admitted), and properties.

We express the descriptive variables in terms of the port variables: thus (Fig. 5.40) we find the voltage across $2R$ and the current in R (Ohm's law), the voltage on port 1 of the gyrator (KVL), and the current in port 2 of the gyrator (gyrator equation). Then, from the second gyrator equation, we find $i_2 = g_m v_1$, which is a first descriptive equation. Moreover, from KCL for cutset 1, we find $i_1 = g_m(2Ri_2 - v_2) + v_1/R = g_m(2Rg_m v_1 - v_2) + v_1/R = v_1 \left(\dfrac{2R^2 g_m^2 + 1}{R} \right) - g_m v_2$, which is the second descriptive equation.

From the descriptive equations, the admitted bases are (i_1, i_2) (corresponding to matrix R), (v_1, v_2) (corresponding to matrix G), and (i_1, v_2) (corresponding to matrix H). Matrices T and T' are also admitted.

The conductance matrix can be directly identified from the descriptive equations:

$$G = \begin{pmatrix} \dfrac{2R^2 g_m^2 + 1}{R} & -g_m \\[2ex] g_m & 0 \end{pmatrix}$$

Properties: The two-port is linear, time-invariant, memoryless, nonreciprocal (provided that $g_m \neq 0$), nonsymmetrical, and passive. This last property

Fig. 5.40 Solution of Case Study 2

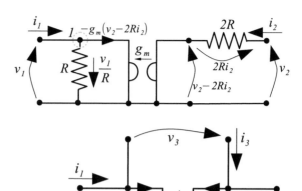

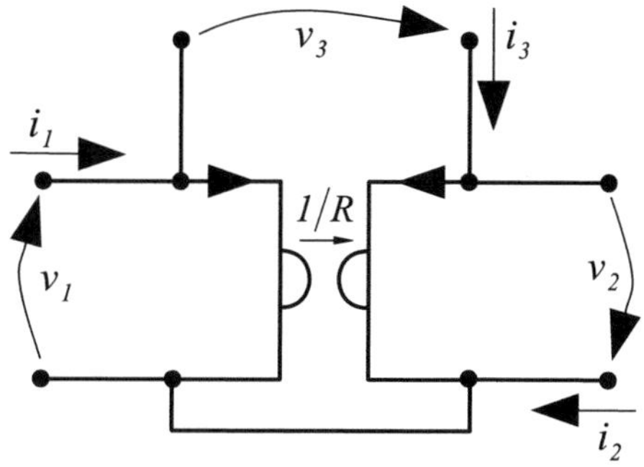

Fig. 5.41 Case Study 3 for gyrator

can be assessed either by computing the absorbed power $p = v_1 i_1 + v_2 i_2 =$

$$v_1^2 \left(\frac{2R^2 g_m^2 + 1}{R} \right) - g_m v_1 v_2 + g_m v_1 v_2 = v_1^2 \left(\frac{2R^2 g_m^2 + 1}{R} \right) \geq 0 \text{ or by taking}$$

into account that the composite two-port contains only passive (the resistors) and nonenergic (the gyrator) components, thus it can only absorb positive power.

Case Study 3

Find the descriptive equations of the composite three-port shown in Fig. 5.41, with $g_m = 1/R$.

We easily find (from topological equations and gyrator descriptive equations) that $v_1 = Ri_2 - Ri_3$, $v_2 = -Ri_1 + Ri_3$, and $v_3 = -v_1 - v_2 = Ri_1 - Ri_2$.

The resistance matrix can be directly identified from the descriptive equations:

$$\begin{pmatrix} 0 & R & -R \\ -R & 0 & R \\ R & -R & 0 \end{pmatrix}$$

This three-port is nonenergic (see the Case Study in Sect. 5.3.5) and is called the *circulator*.

5.8   Connections of Two-Ports

When two or more linear, time-invariant, and memoryless two-port elements are connected, the two-port parameters of the combined network can be found by performing matrix algebra on the matrices of parameters for the component two-ports. The matrix operation can be made particularly simple with an appropriate choice of two-port parameters to match the type of connection of the two-ports, as shown in the following subsections.

The combination rules need to be applied with care. Some connections result in the port condition being invalidated or Kirchhoff's laws violated and the combination rule will no longer apply.

5.8.1   *Cascade Connection*

One of the most common two-port connections is the so-called *cascade connection*.

> In the **cascade connection** two two-ports are connected with port 2 (output port) of the first connected to port 1 (input port) of the second, as shown in Fig. 5.42.

The most convenient representation for cascade-connected two-ports is the forward transmission matrix T. If both two-ports admit transmission matrices T_A and T_B and also the resulting two-port admits transmission matrix T, then T can be easily obtained as $T = T_A T_B$. Indeed, for two-port A, we have

$$\begin{pmatrix} v_1 \\ i_1 \end{pmatrix} = T_A \begin{pmatrix} v_2 \\ -i_2 \end{pmatrix}$$

and similarly, for two-port B, we have

$$\begin{pmatrix} v_2 \\ -i_2 \end{pmatrix} = T_B \begin{pmatrix} v_3 \\ -i_3 \end{pmatrix}$$

Then,

Fig. 5.42 Cascade connection of two-ports

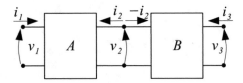

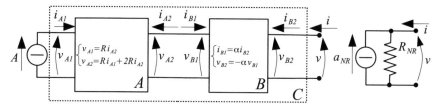

Fig. 5.43 Case Study for cascade connection

$$\begin{pmatrix} v_1 \\ i_1 \end{pmatrix} = T_A T_B \begin{pmatrix} v_3 \\ -i_3 \end{pmatrix}$$

Cascade connections are quite common in electronics, because by connecting multiple amplifiers in this way one can obtain higher gains.

Case Study

Given the composite two-terminal shown in Fig. 5.43, find the matrix representations admitted by two-port A and by two-port B and the Norton equivalent representation shown in the figure.

Two-port A admits all matrix representations but H', because the basis (v_{A1}, i_{A2}) is not admitted.

Two-port B admits all matrix representations but R and G, because current and voltage bases are not admitted.

From the descriptive equations, we directly find $T_A = \begin{pmatrix} 0 & -R \\ \dfrac{1}{R} & 2 \end{pmatrix}$ and

$T_B = \begin{pmatrix} -\dfrac{1}{\alpha} & 0 \\ 0 & -\alpha \end{pmatrix}$, therefore the cascade connection has forward transmission

matrix $T_C = T_A T_B = \begin{pmatrix} 0 & R\alpha \\ -\dfrac{1}{R\alpha} & -2\alpha \end{pmatrix}$ (notice that the reciprocity condition is

fulfilled).

Then, from the more compact equivalent representation shown in Fig. 5.44, we obtain $A = -\dfrac{v}{R\alpha} + 2\alpha i$; that is, $i = \dfrac{v}{2\alpha^2 R} + \dfrac{A}{2\alpha}$. Then, $R_{NR} = 2\alpha^2 R$

and $a_{NR} = -\dfrac{A}{2\alpha}$.

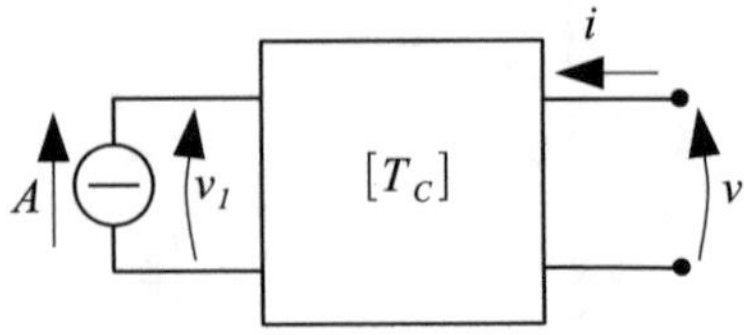

Fig. 5.44 More compact
representation of the
two-terminal shown in
Fig. 5.43

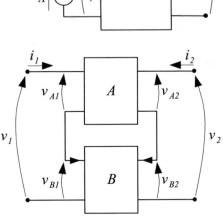

Fig. 5.45 Series connection
of two-ports

5.8.2   Series-Series Connection

Another quite diffused connection is the so-called *series-series connection* (or simply
series connection).

> In the **series connection** two two-ports are connected in such a way that the
> ports 1 of both two-ports share the same descriptive current, as well as the
> ports 2, as shown in Fig. 5.45.

The most convenient representation for series-connected two-ports is the resistance matrix R. If both two-ports admit resistance matrices R_A and R_B and also
the resulting two-port admits resistance matrix R, then R can be easily obtained as
$R = R_A + R_B$. Indeed, for two-port A, we have

$$\begin{pmatrix} v_{A1} \\ v_{A2} \end{pmatrix} = R_A \begin{pmatrix} i_1 \\ i_2 \end{pmatrix}$$

and similarly, for two-port B, we have

$$\begin{pmatrix} v_{B1} \\ v_{B2} \end{pmatrix} = R_B \begin{pmatrix} i_1 \\ i_2 \end{pmatrix}$$

Thus,

$$\begin{pmatrix} v_1 \\ v_2 \end{pmatrix} = \begin{pmatrix} v_{A1} \\ v_{A2} \end{pmatrix} + \begin{pmatrix} v_{B1} \\ v_{B2} \end{pmatrix} = [R_A + R_B] \begin{pmatrix} i_1 \\ i_2 \end{pmatrix}$$

Fig. 5.46 Parallel
connection of two-ports

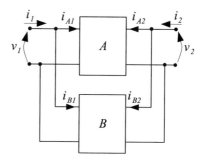

5.8.3 Parallel-Parallel Connection

A third important connection is the so-called *parallel-parallel connection* (or simply *parallel connection*).

> In the **parallel connection** two two-ports are connected in such a way that the ports 1 of both two-ports share the same descriptive voltage, as well as the ports 2, as shown in Fig. 5.46.

The most convenient representation for parallel-connected two-ports is the conductance matrix G. If both two-ports admit conductance matrices G_A and G_B and the resulting two-port admits conductance matrix G, then G can be easily obtained as $G = G_A + G_B$. Indeed, for two-port A, we have

$$\begin{pmatrix} i_{A1} \\ i_{A2} \end{pmatrix} = G_A \begin{pmatrix} v_1 \\ v_2 \end{pmatrix}$$

and similarly, for two-port B, we have

$$\begin{pmatrix} i_{B1} \\ i_{B2} \end{pmatrix} = G_B \begin{pmatrix} v_1 \\ v_2 \end{pmatrix}$$

Then,

$$\begin{pmatrix} i_1 \\ i_2 \end{pmatrix} = \begin{pmatrix} i_{A1} \\ i_{A2} \end{pmatrix} + \begin{pmatrix} i_{B1} \\ i_{B2} \end{pmatrix} = [G_A + G_B] \begin{pmatrix} v_1 \\ v_2 \end{pmatrix}$$

5.8.4 Other Connections and Case Studies

When ports 1 (2) are connected in parallel and ports 2 (1) are connected in series, we have the so-called parallel-series (series-parallel) connection. In these cases, the most convenient representations are the hybrid matrices.

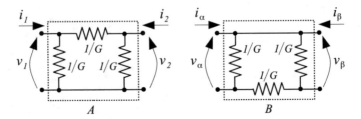

Fig. 5.47 Case Study 1

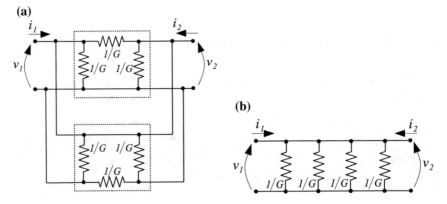

Fig. 5.48 Parallel connection of the two-ports A and B shown in Fig. 5.47 (a) and its equivalent representation (b)

Two case studies summarizing the main concepts for two-port connections follow.

Case Study 1

Compute, if admitted, the conductance matrix for the parallel connection of the two-ports A and B shown in Fig. 5.47.

Two-port A admits the conductance matrix and $G_A = G \begin{pmatrix} 2 & -1 \\ -1 & 2 \end{pmatrix}$. (The two-port is reciprocal and symmetrical.)

Two-port B admits the conductance matrix and $G_B = G_A$.

The parallel connection is shown in Fig. 5.48a and is equivalent to the two-port shown in Fig. 5.48b, which clearly does not admit the voltage basis. Then, the parallel connection does not admit the conductance matrix.

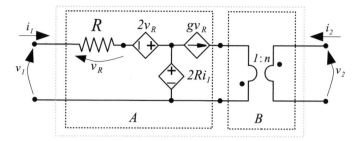

Fig. 5.49 Case Study 2

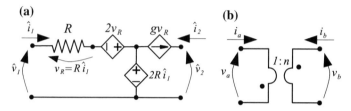

Fig. 5.50 Two-port A (a) and two-port B (b)

Case Study 2

Given the composite two-port shown in Fig. 5.49, find the matrix represen-
tations admitted by two-port A and by two-port B and the transmission matrix
of their cascade connection. Is the series connection admitted? Is the parallel
connection admitted?

Figure 5.50a shows two-port A, whose descriptive equations are $\hat{v}_1 = R\hat{i}_1$
and $\hat{i}_2 = -gR\hat{i}_1$. Then A admits matrices G, H, and $T_A = \dfrac{1}{gR}\begin{pmatrix} 0 & R \\ 0 & 1 \end{pmatrix}$.

Figure 5.50b shows two-port B, whose descriptive equations are $v_a = \dfrac{-v_b}{n}$

and $i_a = ni_b$. Then B admits matrices H, H', T', and $T_B = \begin{pmatrix} -\dfrac{1}{n} & 0 \\ 0 & -n \end{pmatrix}$.

The cascade connection has forward transmission matrix $T = T_A T_B =$
$\begin{pmatrix} 0 & -\dfrac{n}{g} \\ 0 & -\dfrac{n}{Rg} \end{pmatrix}$.

The series connection is not admitted because both two-ports do not admit
the resistance matrix. The parallel connection is not admitted because two-port
B does not admit the conductance matrix.

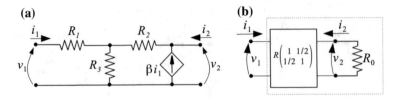

Fig. 5.51 Problems 5.1 (**a**) and 5.2 (**b**)

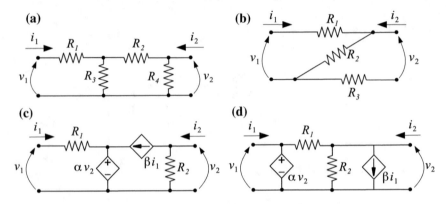

Fig. 5.52 Problem 5.3

5.9 Problems

5.1 Find, if admitted, the matrices R, G, H, and T for the two-port shown in Fig. 5.51a.

5.2 Find the Thévenin equivalent representation shown in Fig. 3.26a for the two-terminal shown in Fig. 5.51b. The two-port is described by the resistance matrix shown in the same figure. Find the value of $R_0 > 0$ such that $R_{TH} = R_0$.

5.3 Find, if admitted, the matrices R, G, H, and T for the two-ports shown in Fig. 5.52 and check their reciprocity and symmetry properties.

5.4 Find admitted bases and matrices for the two-ports shown in Fig. 5.53. If admitted, find out the forward transmission matrix T.

5.5 For the circuits shown in Fig. 5.54, find i (a) and v_o (b, c).

5.6 For the circuit shown in Fig. 5.55, find:

1. The power delivered by voltage source E_1
2. The voltage v_1

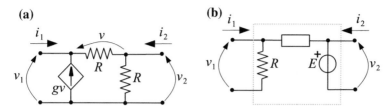

Fig. 5.53 Problem 5.4

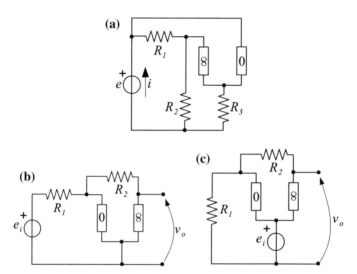

Fig. 5.54 Problem 5.5

Fig. 5.55 Problem 5.6

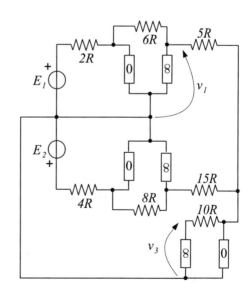

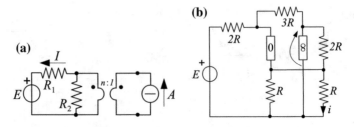

Fig. 5.56 Problems 5.7 (a) and 5.8 (b)

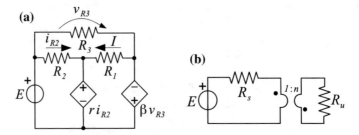

Fig. 5.57 Problems 5.9 (a) and 5.10 (b)

3. The voltage v_3

Find also the numerical solutions for $E_1 = 2$ V, $R = 1$ kΩ, $E_2 = 3$ V.

5.7 For the circuit shown in Fig. 5.56a, find the current I. Also find the numerical solution for $E = 1$ V, $R_1 = 100\Omega$, $R_2 = 400\Omega$, $A = 20$ mA, $n = 2$.

5.8 For the circuit shown in Fig. 5.56b, find:

1. The current i
2. The power absorbed by the nullor

Also find the numerical solutions for $E = 0.5$ V, $R = 100\Omega$.

5.9 For the circuit shown in Fig. 5.57a, find the current I. Also find the numerical solution for $E = 4$ V, $R_1 = 50\Omega$, $R_2 = 15\Omega$, $r = 25\Omega$, $\beta = 3$.

5.10 For the circuit shown in Fig. 5.57b, find the ratio between the power p_u absorbed by R_u and the power p_E delivered by the voltage source.

5.11 For the circuit shown in Fig. 5.58a, where $g > 0$, find:

1. The current i_∞
2. The voltage v_∞
3. The power delivered by R_3

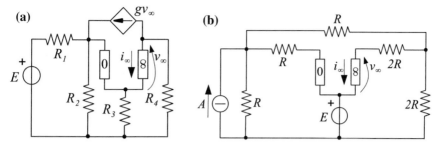

Fig. 5.58 Problems 5.11 (**a**) and 5.12 (**b**)

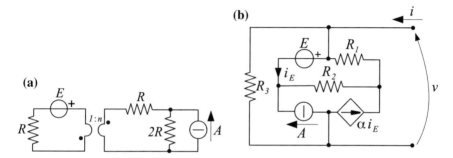

Fig. 5.59 Problems 5.13 (**a**) and 5.14 (**b**)

Also find the numerical solutions for $E = 5$ V, $R_1 = 10\Omega$, $R_2 = 20\Omega$, $R_3 = 10\Omega$, $R_4 = 50\Omega$, $g = 100$ mΩ^{-1}.

5.12 For the circuit shown in Fig. 5.58b, find:

1. The current i_∞
2. The voltage v_∞
3. The power delivered by the voltage source
4. The power absorbed by the current source

5.13 For the circuit shown in Fig. 5.59a, find:

1. The power delivered by the voltage source
2. The power absorbed by the ideal transformer

Also find the numerical solutions for $E = 1$ V, $R = 50\Omega$, $A = 1$ mA, $n = 2$.

5.14 Find the Thévenin equivalent representation shown in Fig. 3.26a for the two-terminal shown in Fig. 5.59b.

5.15 For the circuit shown in Fig. 5.60, where

Fig. 5.60 Problem 5.15

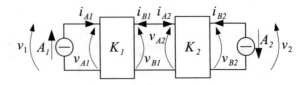

Fig. 5.61 Problem 5.16

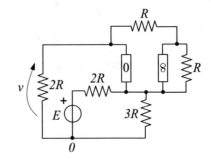

$$\begin{cases} v_{A1} = v_{B1} - Ri_{B1} \\ i_{A1} = gv_{B1} \end{cases} \quad \text{and} \quad \begin{cases} v_{B2} = 2v_{A2} \\ i_{B2} = i_{A2} - gv_{B2} \end{cases}$$

with $R \neq 0$ and $g \neq 0$, find:

1. The bases admitted by the two-port K_1
2. The bases admitted by the two-port K_2
3. The transmission matrix of the cascade connection of K_1 and K_2
4. The voltage v_1
5. The voltage v_2

5.16 For the circuit shown in Fig. 5.61, find:

1. The voltage v
2. The power absorbed by the nullor

Also find the numerical solutions for $E = 4$ V, $R = 25\Omega$.

5.17 For the circuit shown in Fig. 5.62a, where $T = \begin{pmatrix} \alpha & r \\ G & \beta \end{pmatrix}$, with $\alpha, r, G, \beta > 0$, find the Thévenin and Norton equivalent representations shown in the same figure (panels b and c).

5.18 Find the voltage v_A in the circuit shown in Fig. 5.63a.

5.19 For the two-port shown in Fig. 5.63b, find:

1. Descriptive equations
2. Admitted bases and matrices

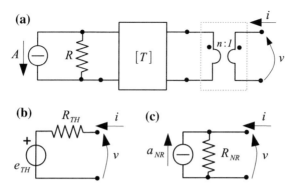

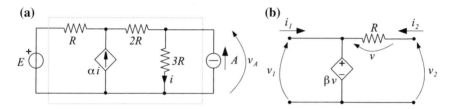

Fig. 5.62 Problem 5.17

Fig. 5.63 Problems 5.18 **(a)** and 5.19 **(b)**

Fig. 5.64 Problem 5.20

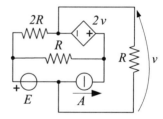

3. Resistance matrix
4. Condition of symmetry (if any) in terms of circuit parameters

5.20 For the circuit shown in Fig. 5.64, find:

1. Voltage v
2. Power delivered by the VCVS

Also find the numerical solutions for $E = 0.6$ V, $R = 30\Omega$, $A = 30$ mA.

5.21 For the composite two-terminal shown in Fig. 5.65, find:

1. The resistance R_{eq} of the resistor equivalent to the composite two-terminal within the dotted grey box

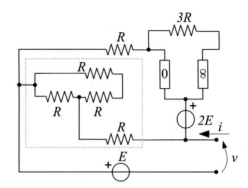

Fig. 5.65 Problem 5.21

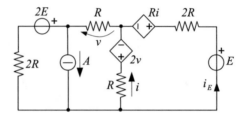

Fig. 5.66 Problem 5.22

2. The Thévenin equivalent representation (Fig. 3.26a)

5.22 For the circuit shown in Fig. 5.66, find:

1. The current i_E
2. The power absorbed by the current source

Also find the numerical solutions for $E = 2$ V, $R = 50\Omega$, $A = 300mA$.

5.23 For the two-ports shown in Fig. 5.67, where $\alpha \neq 0$, find:

1. For the two-port K_1 the admitted bases, the reciprocity condition (if any), and the forward transmission matrix T_1, if admitted.
2. The forward transmission matrix T_2 for the two-port K_2, if admitted.
3. The transmission matrix of the cascade connection of K_1 and K_2.
4. Would the parallel connection of K_1 and K_2 be admitted?
5. Would the series connection of K_1 and K_2 be admitted?

5.24 Find the Norton equivalent representation shown in Fig. 3.30a for the two-terminal shown in Fig. 5.68a.

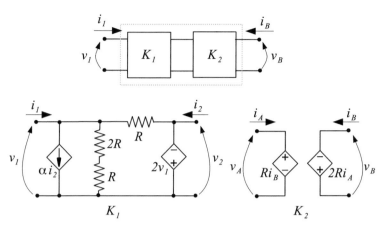

Fig. 5.67 Problem 5.23

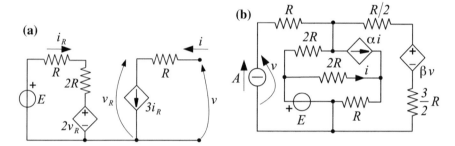

Fig. 5.68 Problems 5.24 (**a**) and 5.25 (**b**)

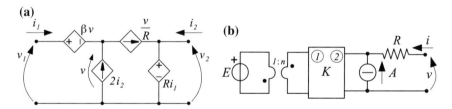

Fig. 5.69 Problem 5.26

5.25 Find the power absorbed by the CCCS in the circuit shown in Fig. 5.68b.

5.26 For the two-port (say K) shown in Fig. 5.69a, where $\beta \neq -1$, find:

1. Descriptive equations
2. Admitted bases
3. Admitted matrices
4. Condition of reciprocity (in terms of parameter β)

Fig. 5.70 Problem 5.27

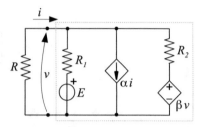

Fig. 5.71 Problem 5.28

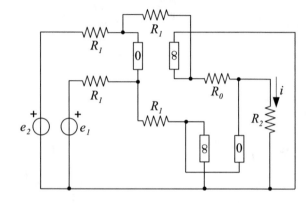

5. Matrix T, if admitted

Now, by imposing the reciprocity condition, find (6.) the Thévenin equivalent representation shown in Fig. 3.26a for the two-terminal shown in Fig. 5.69b.

5.27 For the circuit shown in Fig. 5.70, with $\alpha \neq 1$ and $\beta \neq 1 + R_2/R_1$, find:

1. The Norton equivalent representation (shown in Fig. 3.30a) of the two-terminal within the dashed box
2. The power absorbed by the voltage source, for $\beta = 1, \alpha = 0, R_1 = R$

5.28 For the circuit shown in Fig. 5.71, find:

1. The current i
2. The functionality of this circuit

5.29 Find the current i in the circuit shown in Fig. 5.72.

Fig. 5.72 Problem 5.29

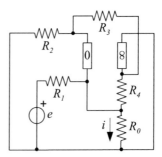

References

1. Premoli A, Storace M (2004) Two-port ideal power transferitors: a unified introduction to ideal transformer and gyrator. IEEE Trans Circuits Syst-II 51:426–429
2. Darlington S (1984) A history of network synthesis and filter theory for circuits composed of resistors, inductors, and capacitors. IEEE Trans Circuits Syst-I 11:1–13
3. Belevitch V (1962) Summary of the history of circuit theory. Proc IRE 50:848–855
4. Foster R (1962) Academic and theoretical aspects of circuit theory. Proc IRE 50:866–871
5. Campbell G, Foster R (1920) Maximum output networks for telephone substation and repeater circuits. Trans Am Inst Electr Eng 39:231–280
6. Tellegen B (1948) The gyrator, a new electric network element. Philips Res Rep 3:81–101
7. Morse A, Huelsman L (1964) A gyrator realisation using operational amplifiers. IEEE Trans Circuit Theory CT-11:277–278
8. Bendik J (1967) Equivalent gyrator networks with nullators and norators. IEEE Trans Circuit Theory CT–14:98
9. Antoniou A (1969) Realisation of gyrators using operational amplifiers, and their use in RC-active-network synthesis. Proc Inst Electr Eng 116:1838–1850

Chapter 6
Advanced Concepts

Abstract In this chapter, Tellegen's theorem and the colored edge theorem are used as tools to get some general properties for a circuit such as reciprocity and passivity, results concerning the power balance, and the so-called no-gain properties. Henceforth, the word *network* is used to denote a generic portion of a circuit, that is, a composite multiport or a composite multiterminal.

6.1 Tellegen's Theorem: A Tool to Investigate Circuit Properties

Tellegen's theorem (Sect. 2.3) can be employed to study the properties of a circuit in terms of those of its components. For this purpose it is useful (although not essential) to introduce a theorem formulation in which the circuit branches containing sources are put in evidence with respect to the rest of the circuit.

Consider a circuit containing voltage and current sources. This circuit can be represented schematically as in Fig. 6.1, where all the P sources are placed outside a closed boundary, whereas the network consisting of all the remaining circuit elements is inside the boundary.

The two-terminal element represented within the boundary does not introduce any restriction on the number of terminals of the inner components. It should be thought of as simply representing any inner edge of the circuit graph. Let N be the number of edges of the graph corresponding to the circuit within the boundary. The descriptive variables for each of these edges follow the standard choice and are denoted by v_k, i_k ($k = 1, \ldots, N$). On the other hand, the conventional directions assumed for the descriptive variables $\hat{v}_j$, $\hat{i}_j$ of the P sources ($j = 1, \ldots, P$) follow the nonstandard choice. With this in mind, the orthogonal vectors of descriptive variables in the circuit are $(\hat{v}_1, \ldots, \hat{v}_P, v_1, \ldots, v_N)$ and $(-\hat{i}_1, \ldots, -\hat{i}_P, i_1, \ldots, i_N)$ and we can write Tellegen's theorem in the form

$$\sum_{j=1}^{P} \hat{v}_j \hat{i}_j = \sum_{k=1}^{N} v_k i_k \qquad (6.1)$$

© Springer International Publishing AG 2018
M. Parodi and M. Storace, *Linear and Nonlinear Circuits:
Basic & Advanced Concepts*, Lecture Notes in Electrical Engineering 441,
DOI 10.1007/978-3-319-61234-8_6

Fig. 6.1 Circuit represented as the connection between P independent sources and a network (a P-port in the figure) containing the remaining elements

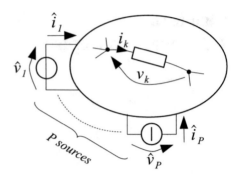

This formulation of Tellegen's theorem can be used to show that *if a network consists entirely of elements of a given class, it belongs in turn to the same class*. Some examples of this general property and other consequences of Tellegen's theorem are provided in the remaining sections of this chapter.

6.2   Passivity

Theorem 6.1 *A network containing only passive elements is in turn passive.*

Proof For a network containing only passive elements, we have

$$\sum_{k=1}^{N} v_k i_k \geq 0,$$

which implies that the whole power entering the network fulfills the same inequality:

$$\sum_{j=1}^{P} \hat{v}_j \hat{i}_j \geq 0;$$

that is, the network is passive.   □

Similarly, when all the network elements are nonenergic, the network is nonenergic too.

6.3   Reciprocity

Theorem 6.2 (Reciprocity theorem) *A network consisting entirely of reciprocal elements is itself reciprocal.*

✂ **Shortcut**. The proof can be skipped without compromising the comprehension of the next sections.

Proof For the circuit of Fig. 6.1, we consider two sets of voltages and currents compatible with the graph; that is, according to Sect. 2.3, they satisfy the KVL and KCL equations, respectively:

$$(\hat{v}'_1, \ldots, \hat{v}'_P, v'_1, \ldots, v'_N) \quad (-\hat{i}'_1, \ldots, -\hat{i}'_P, i'_1, \ldots, i'_N)$$

$$(\hat{v}''_1, \ldots, \hat{v}''_P, v''_1, \ldots, v''_N) \quad (-\hat{i}''_1, \ldots, -\hat{i}''_P, i''_1, \ldots, i''_N)$$

By applying Tellegen's theorem in the form Eq. 6.1 to the cross-powers, we have:

$$\sum_{j=1}^{P} \hat{v}''_j \hat{i}'_j = \sum_{k=1}^{N} v''_k i'_k$$

$$\sum_{j=1}^{P} \hat{v}'_j \hat{i}''_j = \sum_{k=1}^{N} v'_k i''_k$$

Then, subtracting the second equation from the first one, we obtain:

$$\sum_{j=1}^{P} \left[\hat{v}''_j \hat{i}'_j - \hat{v}'_j \hat{i}''_j \right] = \sum_{k=1}^{N} \left[v''_k i'_k - v'_k i''_k \right]$$

All the network components are reciprocal by assumption, therefore the sum at the r.h.s. of the equation is zero, and we have

$$\sum_{j=1}^{P} \hat{v}''_j \hat{i}'_j = \sum_{j=1}^{P} \hat{v}'_j \hat{i}''_j$$

which is precisely the definition of reciprocity for a network. □

Case Study

The resistance matrix R of the three-terminal in the circuit of Fig. 6.2 is

$$R = \begin{bmatrix} R_{11} & R_M \\ R_M & R_{22} \end{bmatrix}.$$

Check the reciprocity of the two-port inside the dashed gray closed boundary.

We notice that the two-port contains only reciprocal elements (matrix R is symmetrical), thus, according to Theorem 6.2, it is reciprocal. The following calculations are just the theorem proof applied to the case study.

Following the reference directions introduced in Fig. 6.2, we consider two sets of voltages and currents compatible with the graph:

$$(\hat{v}'_1, \hat{v}'_2, v'_1, v'_2, v'_3, v'_a, v'_b) \quad (-\hat{i}'_1, -\hat{i}'_2, i'_1, i'_2, i'_3, i'_a, i'_b)$$

$$(\hat{v}''_1, \hat{v}''_2, v''_1, v''_2, v''_3, v''_a, v''_b) \quad (-\hat{i}''_1, -\hat{i}''_2, i''_1, i''_2, i''_3, i''_a, i''_b).$$

By applying Tellegen's theorem in the form 6.1 to the cross-powers, we obtain:

$$\sum_{j=1}^{2} \hat{v}''_j \hat{i}'_j = \sum_{k=1}^{3} v''_k i'_k + v''_a i'_a + v''_b i'_b$$

$$\sum_{j=1}^{2} \hat{v}'_j \hat{i}''_j = \sum_{k=1}^{3} v'_k i''_k + v'_a i''_a + v'_b i''_b.$$

Taking into account the descriptive equations of the network components, the above equations can be recast as

$$\sum_{j=1}^{2} \hat{v}''_j \hat{i}'_j = \sum_{k=1}^{3} R_k i''_k i'_k + (R_{11} i''_a + R_M i''_b) i'_a + (R_M i''_a + R_{22} i''_b) i'_b$$

$$\sum_{j=1}^{2} \hat{v}'_j \hat{i}''_j = \sum_{k=1}^{3} R_k i'_k i''_k + (R_{11} i'_a + R_M i'_b) i''_a + (R_M i'_a + R_{22} i'_b) i''_b.$$

Inasmuch as the right-hand sides of these equations are equal, we have:

$$\sum_{j=1}^{2} \hat{v}''_j \hat{i}'_j = \sum_{j=1}^{2} \hat{v}'_j \hat{i}''_j$$

Fig. 6.2 Case Study

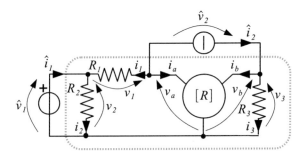

which proves that the network is reciprocal.

6.4 Circuit Biasing and Small-Signal Relations

Many electronic circuits containing nonlinear components, such as communications and signal processing devices, usually carry a small-amplitude AC signal, also called a *small signal*, summed to a DC term, also called *bias*. This suggests that these circuits can be studied by analyzing relatively small perturbations around a bias point: this is usually done by linearizing the circuit equations around the bias point.

As an example, in the circuit shown in Fig. 6.3a the DC voltage E generates the bias voltages $v_1^{(0)}$ and $v_2^{(0)}$ across the resistor and the diode, respectively; let $\hat{i}_1^{(0)}$ be the (common) bias current. All these values can be found by solving the circuit equations. The diode's equation is obviously nonlinear. The introduction of the AC voltage source δe induces current and voltage perturbations with respect to the bias values, as shown in Fig. 6.3c, d.

More in general, it is often possible to decompose a time-varying signal as the sum of a large-amplitude component (large signal) and a small-amplitude component (small signal). The DC and AC signals are just a particular case of this general scenario.

Tellegen's theorem relates the bias and perturbation values without introducing any assumption about the amplitude of the perturbations, which are not necessarily small signals. In the following, we provide two results concerned with power balances for both the biases and the perturbation values of voltages and currents. Making reference to Fig. 6.1, we first consider a vector $v^{(0)}$ and a vector $i^{(0)}$ whose voltage and current elements

$$v^{(0)} = (\hat{v}_1^{(0)}, \ldots, \hat{v}_P^{(0)}, v_1^{(0)}, \ldots, v_N^{(0)})$$

$$i^{(0)} = (-\hat{i}_1^{(0)}, \ldots, -\hat{i}_P^{(0)}, i_1^{(0)}, \ldots, i_N^{(0)})$$

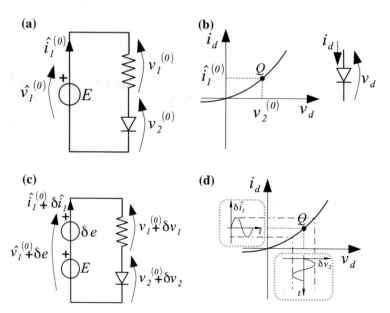

Fig. 6.3 Nonlinear circuit: bias voltages and currents (**a**); bias point Q for the diode (**b**); perturbed voltages and currents (**c**); small signals for the diode (**d**)

Fig. 6.4 The pairs of vectors to which Tellegen's theorem is applied

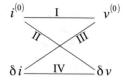

are such that all circuit equations are fulfilled. They correspond to a physical reference situation for the circuit and can be either constant values (bias values) or time-dependent functions (e.g., sinusoidal functions with assigned frequency). Obviously, $v^{(0)}$ and $i^{(0)}$ are orthogonal. Now consider a situation in which the voltages and currents are perturbed with respect to these reference values. This means that the voltage and current vectors are now structured as follows.

$$v = v^{(0)} + \delta v$$

$$i = i^{(0)} + \delta i. \tag{6.2}$$

Voltages and currents of the vectors v and i are compatible with the graph of the circuit and orthogonal as well. From expressions Eq. 6.2, then, it follows that the perturbation vectors δv and δi, individually, also share the same properties.

Tellegen's theorem in the form 6.1 can be applied to the elements of the vectors $v^{(0)}$, $i^{(0)}$, δv, and δi defined above in four ways, labeled in the diagram of Fig. 6.4.

The terms that appear in the cross-power balances II and III have a mixed character. Each edge of the graph contributes to the Tellegen balance with a product term between one of its descriptive variables in the reference state and its complementary variable in the perturbed state. As such, they do not, in general, offer a direct physical interpretation.

The other two cross-power balances are explicitly written below:

$$\sum_{j=1}^{P} \hat{v}_j^{(0)} \hat{i}_j^{(0)} = \sum_{k=1}^{N} v_k^{(0)} i_k^{(0)} \tag{6.3}$$

$$\sum_{j=1}^{P} \delta\hat{v}_j \delta\hat{i}_j = \sum_{k=1}^{N} \delta v_k \delta i_k \tag{6.4}$$

Equation 6.3 is simply the balance of powers in the reference physical situation: the overall power supplied by the sources is absorbed by the components inside the boundary surface.

Equation 6.4 is concerned with the power balancing related to perturbations. The power terms for these components may evolve with time differently from those in the reference physical situation. For example, if the reference terms in Eq. 6.3 are constants, the perturbation terms in Eq. 6.4 may be periodic in time (e.g., sinusoidal, as in the examples cited at the beginning of this section). Alternatively, if the reference terms are large AC signals at a certain frequency, the perturbation terms may be, say small AC signals with a higher frequency.

The result of Eq. 6.4 is a nontrivial consequence of Tellegen's theorem and is the basis for the properties described in the next section.

6.5 Local Activity and Amplification

When all the N (with $N \geq 2$) products $\delta v_k \delta i_k$ in the sum at the r.h.s. of Eq. 6.4 are positive, each of them is strictly smaller than the overall power $\sum_{j=1}^{P} \delta\hat{v}_j \delta\hat{i}_j$ delivered by the sources. However, there are cases where one or more of these products can be negative. As an example, for a nonlinear resistor whose DP characteristic on the (i, v) plane has a negative slope $-R$ at its bias point Q (Fig. 6.5), the small-signal variations δv and δi are such that $\delta v \delta i = -R(\delta i)^2 < 0$. In this case, we say that the resistor is *locally active* at its bias point.

A resistor is said to be **locally active** at a specific point of its DP characteristic if the slope of the characteristic is negative at that point.

Fig. 6.5 DP characteristic of a resistor locally active at Q. In a small neighborhood of Q (*grey circle*) $\delta v \simeq -R\delta i$

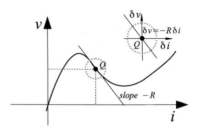

In the presence of a locally active resistor, we can get an amplification. The following particular case of the power balance expressed by Eq. 6.4 provides an example, with one source ($P = 1$) and two resistors inside the boundary ($N = 2$). In this case Eq. 6.4 becomes:

$$\underbrace{\delta \hat{v} \delta \hat{i}}_{p_s} = \underbrace{\delta v_1 \delta i_1}_{p_1} + \underbrace{\delta v_2 \delta i_2}_{p_2} .$$

The term p_s, positive by assumption, denotes the small signal power supplied by the external source and entering the boundary surface, whereas p_1 and p_2 are the powers absorbed by the inner resistors. Taking account of this expression, the ratio between the powers p_1 and p_s can be recast as

$$\frac{p_1}{p_s} = \frac{p_s - p_2}{p_s} .$$

- When $p_1 > 0$ and $p_2 > 0$ we have

$$\frac{p_1}{p_s} < 1 \quad \text{(no amplification)}$$

A similar result holds for $\frac{p_2}{p_s}$.

- If $p_1 > 0$ and $p_2 < 0$, instead, we obtain:

$$\frac{p_1}{p_s} > 1.$$

Therefore the power p_1 is larger than the power p_s entering the boundary surface; that is, there is an amplification of the power related to (small) perturbations. The resistor with $p_2 < 0$ is *locally active*.

We remark that these results relate only to electrical perturbations with respect to a reference physical situation.

To better understand the roles of the bias and of the perturbation terms, consider the following case study.

Case Study

Figure 6.6a shows the physical reference situation. A constant voltage source E is connected to the series of a linear resistor R_o and of a nonlinear, passive, current-controlled resistor whose DP characteristic $v = \bar{v}(i)$ is shown in Fig. 6.6b. In terms of the current i_o flowing through the circuit elements, we can write:

$$E = R_o i_o + \bar{v}(i_o). \tag{6.5}$$

We assume that, owing to the parameters' setting, at the point $Q = (i_o, \bar{v}(i_o))$ the resistor characteristic has a negative slope $-R$ (with $R > 0$). (See Fig. 6.6b.)

The voltage source δe (Fig. 6.7) perturbs the current and voltages with respect to the reference values. In this case Eq. 6.5 is recast as

$$E + \delta e = R_o(i_o + \delta i) + \bar{v}(i_o + \delta i) \tag{6.6}$$

Assuming that, in magnitude, perturbations are smaller than the corresponding reference values (in particular, $|\delta i| << |i_o|$), the nonlinear resistor's voltage can be approximated by a Taylor series centered at the bias point Q and truncated to the first term:

$$\bar{v}(i_o + \delta i) \simeq \bar{v}(i_o) + \underbrace{\frac{d\bar{v}}{di}\Big|_{i_o}}_{-R} \delta i.$$

Then, Eq. 6.6 can be approximated:

$$E + \delta e \simeq R_o(i_o + \delta i) + \bar{v}(i_o) - R\delta i$$

from which, taking account of Eq. 6.5, and within the limits of accuracy of the Taylor first-order approximation, we obtain

$$\delta e = (R_o - R)\delta i. \tag{6.7}$$

This result allows us to write Eq. 6.4 as

$$\delta e \delta i = R_o(\delta i)^2 - R(\delta i)^2$$

When $R < Ro$, δi, and δe share the same sign (Eq. 6.7) and the power term $\delta e \delta i$ is positive. Under this assumption, the ratio between the power absorbed by R_o and the power delivered by the source δe is:

(a)   **(b)**

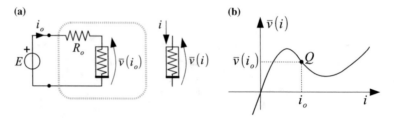

Fig. 6.6 Small-power amplification example: **a** circuit in the reference (bias) situation; **b** DP characteristic $v = \bar{v}(i)$ of the nonlinear resistor

Fig. 6.7 Circuit in the perturbed (small-signal) situation

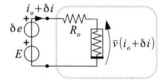

$$\frac{R_o(\delta i)^2}{\delta e \delta i} = \frac{\delta e \delta i + R(\delta i)^2}{\delta e \delta i} = 1 + \underbrace{\frac{R \delta i}{\delta e}}_{>0} > 1.$$

This means that the input small-signal power is amplified, thanks to the power contribution $R(\delta i)^2$ that the locally active nonlinear resistor delivers to R_o.

6.6 Colored Edge Theorem

The colored edge theorem, or colored branch theorem [1], was initially formulated and applied mainly in the graph-theoretic framework. Later on, it was progressively recognized as a powerful tool also in the circuit theory context [2]. The theorem depends on the circuit's graph only, as well as Tellegen's theorem, and has the same level of generality. Here, it is used to prove two theorems in Sect. 6.7.

✂ **Shortcut.** The reader not interested in the cited proofs can skip this section.

For an arbitrary allocation of the edges of a graph in three sets (colors), the theorem establishes alternatives for the existence of loops or cut-sets having specific characteristics (color and orientation of the edges). Following [2], first consider a directed graph; then freely color each of its edges using one of three colors: red, blue, or green. Among the green edges, one is dark. In this manner, the same graph can be colored in many different ways.

In a generic loop inside the graph, the edges oriented in the same direction (clockwise/counterclockwise) are called *similarly directed*. Likewise, considering a cut-set

Fig. 6.8 Examples for the
colored edge theorem. **a**
Loop fulfilling condition 1;
b cut-set fulfilling
condition 2

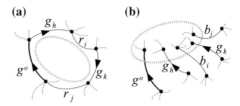

with its associated closed path, we call similarly directed the cut-set edges oriented
as the closed path (or surface for nonplanar graphs).

Theorem 6.3 (Colored edge theorem) *In a directed colored graph (not necessarily
connected), one and only one of the following properties must hold:*

1. *The dark green edge forms a loop exclusively with green and/or red edges such
   that the green edges (including the dark green edge) are similarly directed.*
2. *The dark green edge forms a cut-set exclusively with green and/or blue edges such
   that the green edges (including the dark green edge) in the cut-set are similarly
   directed.*

The proof of the theorem can be found in [2] and is not reported here.

Two examples are shown in Fig. 6.8, where g^o denotes the dark green edge. Each
green edge (g-edges in the figure) is oriented, whereas for red (r-edges of panel a)
and blue (b-edges of panel b) edges, both directions are allowed.

Remark 1   Only one of the two conditions holds, thus the existence of a loop fulfilling
condition 1 of the theorem excludes the existence of a cut-set with the characteristics
given in condition 2; on the other hand, in the presence of a cut-set fulfilling condition
2, we cannot have a loop as described in condition 1.

Remark 2   The theorem says nothing about uniqueness. If a loop (cut-set) fulfills
condition 1 (2), it is possible that the graph contains other loops (cut-sets) with the
same characteristics.

Figure 6.9 shows two possible applications of the theorem, corresponding to two
different colorings of the edges on the same (nonplanar) graph. The choice of the
dark green edge g^o as in Fig. 6.9a leads to a loop formed by another green edge
oriented as g^o and by two red edges. It is easy to verify that g^o does not form cut-sets
meeting condition 2 of the theorem. The same graph is represented in Fig. 6.9b with
a different choice of the edge colors. In this case, the edge chosen as g^o forms a
cut-set with two green edges both oriented as g^o and three blue edges. You can easily
check that g^o does not form loops meeting condition 1 of the theorem. To check
your comprehension, you can arbitrarily choose other color allocations for the graph
edges and verify the theorem validity.

The colored edge theorem is one of the most appropriate results to highlight
the generality of graph theory. Under rather mild assumptions, a graph can be used

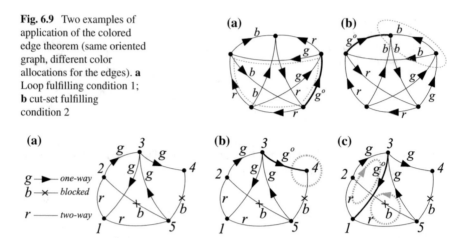

Fig. 6.9 Two examples of application of the colored edge theorem (same oriented graph, different color allocations for the edges). **a** Loop fulfilling condition 1; **b** cut-set fulfilling condition 2

Fig. 6.10 Example of graph representing a vehicular traffic network; **a** graph structure with the various kinds of streets (one-way, two-way, blocked); **b**, **c** results from the colored edge theorem for two different choices of the *dark green edge* g^o

to represent lumped physical systems of very different nature (e.g., mechanical, acoustic, hydraulic, thermal) [3]. In these cases, the voltage and current variables are replaced by equivalent variables (resp., across variables and through variables) suitable to closely represent the system physical nature. The graph plays, in any case, the same role as in the case of circuits; that is, it codes information on how the various parts of the system are connected to each other. Vehicular traffic networks are among these systems [4] and the results obtained by applying the colored edge theorem to their graphs are simple and meaningful.

A road network can be schematized in a completely intuitive way as an interconnection of edges, each representing a street. We can color two-way streets in red, one-way streets in green, and blocked streets in blue. (See Fig. 6.10a.) The edge g^o chosen in Fig. 6.10b, connecting node 3 to node 4, forms a cut-set with a blue edge (and, according to the theorem, no loop with similarly directed g edges and/or with r edges). Then node 4 is reachable from node 3 but not vice versa, owing to the edge orientation. The edge g^o chosen in Fig. 6.10c forms two loops with similarly directed green edges and with red edges (and, according to the theorem, no cut-set with similarly directed g edges and/or b edges): all nodes of these loops can be reached from any of their nodes.

In the circuit field, the colored edge theorem can be a very efficient analysis tool when the edges of the graph are colored depending on the nature (or on a specific property) of the components corresponding to these edges. For instance, consider the diode Graetz bridge in the circuit of Fig. 6.11. In general, the voltage source e on the input side **AB** is time-dependent: $e = e(t)$ (e.g., an AC source as in the case of an electric socket). The output variable is the current $i(t)$ through the resistor R. The

Fig. 6.11 The Graetz circuit

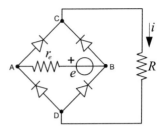

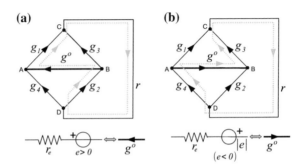

Fig. 6.12 Graphs chosen for the Graetz circuit. All edges are colored in *green* except the output edge CD, colored in *red*. The conventions adopted for the input edge g^o depend on the sign of e, as shown in the lower part of the figure: **a** when $e > 0$; **b** when $e < 0$

four diodes of the bridge are thought to behave as an open or short circuit, according to the PWL DP characteristic shown in Fig. 3.13c.

In the circuit graph, each diode edge is colored in green and oriented as the current direction when the diode operates as a short circuit. The edge g^o is associated with the series connection of $e(t)$ and r_e and is oriented according to the physical direction of the current delivered by the input side AB to the rest of the circuit: from right to left when $e > 0$ (Fig. 6.12a) and from left to right when $e < 0$ (Fig. 6.12b). Finally, the output side CD of the graph is colored in red. Following the colored edge theorem, you immediately see that in both graphs no cut-sets of g^o with similarly directed green edges can occur (condition 2). In both cases, then, g^o forms a loop with similarly directed green edges and with the red edge (condition 1). These loops are marked by the grey dashed lines in Fig. 6.12a, b.

The positive current assumption through the edges g_1, g_2 in the first case and g_3, g_4 in the latter case implies that the corresponding PWL diodes behave as short circuits. In both cases, therefore, the Graetz circuit is equivalent to the simpler circuit shown in Fig. 6.13a, and we have:

$$i = \frac{|e|}{R + r_e}.$$

This proves that, irrespective of the sign of $e(t)$, the current i is "one-directional". We finally observe that it is easy to verify that the diodes on the edges g_3 and g_4 in Fig. 6.12a and the diodes on the edges g_1 and g_2 in Fig. 6.12b behave as open circuits.

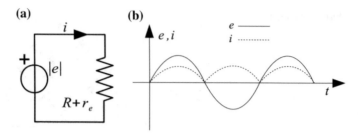

Fig. 6.13 Equivalent Graetz circuit (**a**); sinusoidal input $e(t)$ with the corresponding output $i(t)$ (**b**)

An application of practical interest is when $e(t)$ is an AC input and one wants to obtain a one-directional current (Fig. 6.13b). In this case the circuit is also known as a full-wave bridge rectifier.

6.7   Circuits Consisting of Two-Terminal Elements: No-Gain Theorems

We consider a circuit containing only independent voltage and current sources, two-terminal resistors, and short-circuit and open-circuit elements (Fig. 6.14). Each resistor can be either linear or nonlinear and is assumed to be *strictly passive*, according to the following definition.

> A two-terminal resistor is **strictly passive** if and only if $vi > 0$ for any point (v, i) of its DP characteristic, except the origin $(0, 0)$.

Then a linear resistor with $R > 0$ is strictly passive and the same holds for any resistor whose characteristic lies in the first and third quadrants and intersects the v and i axes only in the origin. For instance, the DP characteristic shown in Fig. 6.6b is strictly passive, whereas the DP characteristic shown in Fig. 3.13c is passive, but not strictly passive.

Furthermore, we assume that the circuit does not contain either loops of short circuits only or cut-sets of open circuits only.

Under these assumptions, the following no-gain theorems hold [2].

Theorem 6.4 *For any element of the circuit, the upper limit of the current magnitude is given by the sum of the current magnitudes through all independent voltage and current sources.*

✂ **Shortcut.** The proof can be skipped without compromising the comprehension of the next sections.

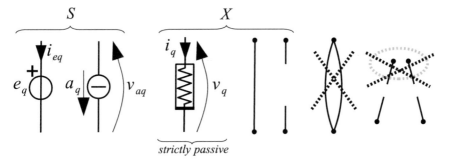

Fig. 6.14 Sets S and X of two-terminal elements allowed by the no-gain theorems. Resistors can be either linear or nonlinear. Loops of short-circuit elements and cut-sets of open-circuit elements are not allowed

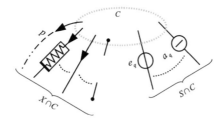

Fig. 6.15 The cutset C in the proof of Theorem 6.4. The conventions for the sources are left unspecified because they are not relevant to the theorem's proof

Proof We preliminarily note that, when the edge considered (henceforth denoted as p) is a voltage or current source, or if $i_p = 0$, Theorem 6.4 is trivially satisfied.

Now consider the case when p is a resistor, or a short circuit or an open circuit and $v_p \neq 0$ or $i_p \neq 0$. The edges of the circuit with zero voltage and current may be removed because they do not contribute to the sum in the theorem's statement, do not alter the circuit solution, and do not introduce new loops, owing to the assumptions. In the resulting circuit, let S be the set of sources, and X the set of resistors, short circuits, and open circuits (Fig. 6.14). Therefore, p is an element of X. Because all of the resistors are strictly passive, without loss of generality we can choose the associated directions for any element (associated with the graph edge q) of X such that $v_q \geq 0$ and $i_q \geq 0$. Now we apply the colored edge theorem by coloring in green the edges in X and in blue the edges in S.

The edge p cannot form a similarly directed loop that contains only other edges of X. Indeed, if $v_p > 0$, the KVL in the loop would be violated because the other loop voltages can only be ≥ 0; if $v_p = 0$, the KVL also implies that the other loop voltages are zero. In the latter case, however, inasmuch as it is excluded that the corresponding currents are also zero (the branches with zero voltage and current have been removed), the edges of the loop should be short circuits, but this is not possible by assumption. According to the colored edge theorem, then, there must be a cut-set C that contains, apart from p, edges from both X and S and wherein the edges from X are similarly directed. (See Fig. 6.15.)

The KCL for cut-set C gives

$$\sum_{q\in(X\cap C)} i_q + \sum_{q\in(S\cap C)} \delta_q i_q = 0 \quad \text{with} \quad \begin{cases} i_q \geq 0 \text{ for } q \in X \cap C \\ \delta_q = \pm 1 \quad \text{(according to the chosen associated directions)} \end{cases}$$

then

$$\left|i_p\right| \equiv i_p \leq \sum_{q\in(X\cap C)} i_q = -\sum_{q\in(S\cap C)} \delta_q i_q \leq \sum_{q\in(S\cap C)} \left|i_q\right| \leq \sum_{q\in S}\left|i_{eq}\right| + \sum_{q\in S}\left|a_q\right|$$

that is,

$$\left|i_p\right| \leq \sum_{q\in S}\left|i_{eq}\right| + \sum_{q\in S}\left|a_q\right|$$

as stated by Theorem 6.4. □

Theorem 6.5 *For any element of the circuit, the upper limit of the voltage magnitude is given by the sum of the voltage magnitudes across all independent voltage and current sources.*

✂ **Shortcut**. The proof can be skipped without compromising the comprehension of the next sections.

Proof We preliminarily note that, when the edge considered (henceforth denoted as p) is a voltage or current source, or if $v_p = 0$, Theorem 6.5 is trivially satisfied.

We again apply the colored edge theorem. We maintain the green color for the edges of X (including the edge p) and red color for the edges of S (no blue edges). As in the proof of Theorem 6.4, the associated directions for all edges in X are chosen so that $v_q \geq 0$ and $i_q \geq 0$. The same line of reasoning, *mutatis mutandis*, immediately gives that edge p cannot form a similarly directed cut-set that contains only edges of X. Therefore, according to the colored edge theorem, there must exist a loop L containing, in addition to the edge p, edges from both sets X and S, where the edges from X are similarly directed. (See Fig. 6.16.) The KCL written for loop L gives

$$\sum_{q\in(X\cap L)} v_q + \sum_{q\in(S\cap L)} \delta_q v_q = 0 \quad \text{with} \quad \begin{cases} v_q \geq 0 \text{ for } q \in X \cap L \\ \delta_q = \pm 1 \end{cases}$$

then

$$\left|v_p\right| \equiv v_p \leq \sum_{q\in(X\cap L)} v_q = -\sum_{q\in(S\cap L)} \delta_q v_q \leq \sum_{q\in(S\cap L)} \left|v_q\right| \leq \sum_{q\in S}\left|e_q\right| + \sum_{q\in S}\left|v_{aq}\right|.$$

that is,

$$\left|v_p\right| \leq \sum_{q\in S}\left|e_q\right| + \sum_{q\in S}\left|v_{aq}\right|$$

as stated by Theorem 6.5. □

Fig. 6.16 The loop L in the proof of Theorem 6.5. The conventions for the sources are left unspecified because they are not relevant to the theorem's proof

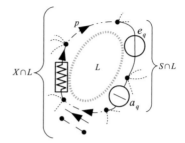

As a particular case for the two theorems, we observe that in a circuit containing a single (voltage or current) source and only strictly passive (both linear and non-linear) resistors, the magnitude of the voltage across any resistor cannot exceed the magnitude of the voltage across the source. A completely analogous statement can be formulated in terms of the current magnitudes through the resistors and through the source, respectively.

References

1. Minty GJ (1960) Monotone networks. Proc R Soc Lond A 257:194–212
2. Vandewalle J, Chua LO (1980) The colored branch theorem and its applications in circuit theory. IEEE Trans Circuits Syst 27:816–825
3. Trent HM (1955) Isomorphisms between oriented linear graphs and lumped physical systems. J Acoustical Soc Am 27:500–527
4. Minty GJ (1961) Solving steady-state nonlinear networks of monotone elements. IRE Trans Circuit Theory 8:99–104

Part IV
Analysis of Memoryless Circuits

Chapter 7
Basic Concepts

Education is the kindling of a flame, not the filling of a vessel.
Socrates

Abstract This chapter is devoted to general methods of circuit analysis, that is, methods to solve a circuit (see Sect. 3.1) in a systematic way. Some of these methods are particularly suitable for circuit simulation using a computer. In particular, here we focus on memoryless circuits, even if the proposed results are easily extended to other circuits in the next chapters. We describe node and mesh analysis methods, the superposition principle, substitution principle, and some practical rules that can help circuit analysis.

7.1 Nodal Analysis

In this section we introduce different versions of the *nodal analysis method*, starting from the simplest one.

7.1.1 Pure Nodal Analysis

We consider a circuit consisting of two-terminals only. Let L be the number of these elements and N the number of circuit nodes. Any one of these nodes can be chosen as the *reference node* (or datum node). We label the reference node with 0 and all the other nodes with integer positive numbers, for example, increasing from 1 to $N - 1$. The voltage e_k from the reference node to the generic node k is called the *node voltage*. It is easy to check (by applying KVL) that any other voltage can be expressed as a sum (with proper signs) of node voltages. This is true, in particular, for the voltages across the components. As an example, for the circuit in Fig. 7.1 the

© Springer International Publishing AG 2018   185
M. Parodi and M. Storace, *Linear and Nonlinear Circuits:*
Basic & Advanced Concepts, Lecture Notes in Electrical Engineering 441,
DOI 10.1007/978-3-319-61234-8_7

Fig. 7.1 Example showing
voltages across components
and node voltages

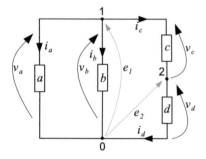

voltages v_a, v_b, v_c, v_d across the components can be expressed in terms of the node
voltages e_1, e_2 as

$$v_a = e_1; \qquad v_c = e_1 - e_2;$$

$$v_b = e_1; \qquad v_d = e_2.$$

The KCL equations are written for all the $N - 1$ nodes different from the reference
node. For the circuit of Fig. 7.1 we have:

$$i_a + i_b + i_c = 0;$$

$$-i_c + i_d = 0.$$

Assumptions: The circuit consists only of

- Linear two-terminal components that admit the voltage basis
- Independent current sources

Under these assumptions, the currents in the KCL equations can be easily
expressed in terms of the node voltages e_k through the component equations. This
leads us to obtain a set of linear equations whose only unknowns are the node voltages e_k ($k = 1, \ldots, N - 1$). Once the node voltages are known, any other voltage or
current can be obtained in a straightforward way. The procedure is detailed through
the following example.

Fig. 7.2 Case Study

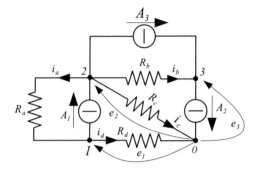

Case Study

Find the symbolic expression of the node voltages in the circuit shown in Fig. 7.2.

Figure 7.2 shows node labeling, node voltages, and currents through the linear resistors; for each component, the voltages are oriented according to the standard choice.

The node equations are:

$$-i_a + i_d + A_1 = 0;$$

$$i_a + i_b + i_c - A_1 + A_3 = 0;$$

$$-i_b + A_2 - A_3 = 0.$$

Each resistor current can now be expressed in terms of the node voltages, for example, $i_a = (e_2 - e_1)/R_a$. After replacing these expressions in the node equations we obtain:

$$\begin{cases} -\frac{1}{R_a}(e_2 - e_1) + \frac{e_1}{R_d} + A_1 = 0 \\ \frac{1}{R_a}(e_2 - e_1) + \frac{1}{R_b}(e_2 - e_3) + \frac{e_2}{R_c} - A_1 + A_3 = 0 \\ -\frac{1}{R_b}(e_2 - e_3) + A_2 - A_3 = 0. \end{cases}$$

After a few manipulations, the system can be recast as

$$\begin{cases} \left(\frac{1}{R_a} + \frac{1}{R_d}\right)e_1 - \frac{1}{R_a}e_2 = -A_1 \\ -\frac{1}{R_a}e_1 + \left(\frac{1}{R_a} + \frac{1}{R_b} + \frac{1}{R_c}\right)e_2 - \frac{1}{R_b}e_3 = A_1 - A_3 \\ -\frac{1}{R_b}e_2 + \frac{1}{R_b}e_3 = -A_2 + A_3. \end{cases}$$

Therefore, the system of equations can be written in compact form:

Fig. 7.3 Equivalent circuit for the Case Study

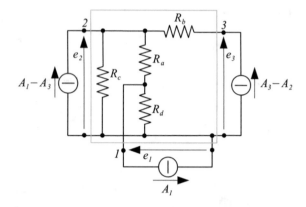

$$Ge = \hat{a}$$

with

$$e = \begin{pmatrix} e_1 \\ e_2 \\ e_3 \end{pmatrix}; \; \hat{a} = \begin{pmatrix} -A_1 \\ A_1 - A_3 \\ -A_2 + A_3 \end{pmatrix}; \; G = \begin{pmatrix} \frac{1}{R_a} + \frac{1}{R_d} & -\frac{1}{R_a} & 0 \\ -\frac{1}{R_a} & \left(\frac{1}{R_a} + \frac{1}{R_b} + \frac{1}{R_c}\right) & -\frac{1}{R_b} \\ 0 & -\frac{1}{R_b} & \frac{1}{R_b} \end{pmatrix},$$

namely in such a way that the terms on the main diagonal of G appear with a positive sign.

In passing, we note that the circuit equations, per se, could be those of a three-port (containing only resistors, which are reciprocal components) described by the matrix G connected to three equivalent current sources, described by the vector $\hat{a}$. (See the equivalent circuit shown in Fig. 7.3.) The three-port is reciprocal, owing to the reciprocity Theorem 3.1. (See also Sect. 6.3.) This implies, for a generalization of the property shown in Sect. 5.4.1 for two-ports, that G is symmetric.

The node voltages can be easily obtained from the system of equations, whatever the form in which the system is written. In any case, the form $Ge = \hat{a}$ is simple to obtain by following a few systematic rules. These rules are easily obtained by observing the correspondences between the terms of the equations and the circuit. Their character, however, is general, and as such they are proposed below.

In the system of equations

$$Ge = \hat{a} \tag{7.1}$$

the entries of G and $\hat{a}$ are obtained by circuit inspection according to the rules:

1. In the ith row, associated with the ith node, the element G_{ii} of the main diagonal is the sum of the conductances of the resistors incident at the node i.

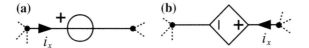

Fig. 7.4 Auxiliary unknown i_x to extend nodal analysis to: **a** independent voltage source; **b** controlled voltage source

2. G_{ik}, with $k \neq i$, is the sum of the conductances (with $-$ sign) of the resistors connecting node i with node k. If no resistors connect node i with node k, then $G_{ik} = 0$.
3. The ith entry of $\hat{a}$ is the algebraic sum of the currents impressed by the sources connected to the ith node; each of these currents contributes with the $+$ sign if its direction is towards the node; otherwise it contributes with the $-$ sign.

According to these rules, everything works as if at the ith node only one current source were present. The source is connected between the ith node and the common node. For instance, in the circuit of Fig. 7.2, the equivalent current source entering node 3 is $A_3 - A_2$.

7.1.2 Modified Nodal Analysis

The nodal analysis method can be modified so as to admit the presence of both independent voltage sources and the four controlled sources defined in Sect. 5.2.1. In this case, the unknown variables of the circuit are not only the node voltages $\{e_k\}$ (*primary unknowns*) but also proper *auxiliary unknowns*.

A first type of auxiliary unknown is related to the voltage sources, both independent and controlled. As a matter of fact, in both cases the component's equation does not provide information about the current through the source. This current must therefore be regarded as an unknown i_x (Fig. 7.4), to be found together with the node voltages.

The second kind of auxiliary unknown is the controlling current for both CCVSs and CCCSs. This descriptive variable appears in the component's equation and must be added to the set of unknown variables. In the case of VCVS and VCCS, instead, because the driving variable is a voltage, it can be directly expressed in terms of node voltages, without the need of an auxiliary unknown.

Because of the introduction of the auxiliary unknowns, the system to be solved no longer has the simple form Eq. 7.1 as for the case of pure nodal analysis.

Many circuit simulation programs, such as those of the SPICE (Simulation Program with Integrated Circuit Emphasis) family, are based on modified nodal analysis (MNA).

Fig. 7.5 Case Study for
MNA

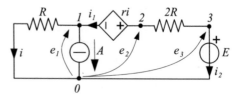

7.1.3 Substitution Rule

Writing the circuit equations can sometimes be made easier by applying the substitution rule described here, which can be used for both independent and controlled sources.

Figure 7.6a shows the portion of a circuit containing a current source A. The branch containing this current source can be removed by modifying the circuit as shown in Fig. 7.6b, where two current sources identical to the original one are connected in parallel to the components a and b. The electrical behavior of the circuit does not change, because the KCL equations that are written at the nodes 1, 2, 3 in the original circuit are exactly the same as for the modified circuit (in particular, the overall contribution of the two current sources at node 1 is zero). As a result of this

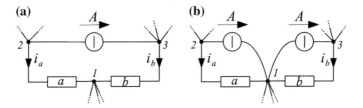

Fig. 7.6 Removing a branch containing a current source: **a** original circuit; **b** equivalent circuit

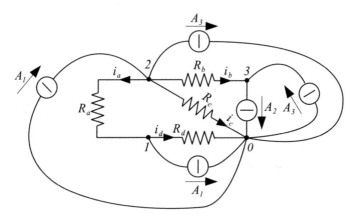

Fig. 7.7 The circuit of Fig. 7.2 after the application of the substitution rule

substitution, we have a current source connected in parallel to both a and b. If the two-terminal a is, in turn, a current source, the parallel connection is equivalent to a single source whose current is given by the algebraic sum of the two constituents. (See Sect. 3.5.2.1.) If a is a resistor, the connection of a current source in parallel originates a Norton equivalent.

As an example, the circuit of Fig. 7.2 can be changed by applying the transformation twice, for both A_1 and A_3. The resulting circuit is shown in Fig. 7.7. The equivalent current sources entering nodes 1, 2, 3 are $-A_1$, $A_1 - A_3$, $-A_2 + A_3$, respectively. Obviously, these are also the elements of the vector $\hat{a}$ obtained from the circuit equations in Sect. 7.1.1.

7.2 Mesh Analysis

Also in this case, there are different versions of the *mesh analysis method*, that require different assumptions. In this chapter, the term mesh is used to denote inner loops only. (See Sect. 2.1.2.)

Fig. 7.8 A planar circuit and
its three meshes

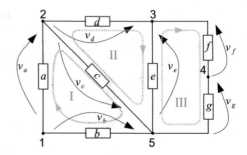

7.2.1 Pure Mesh Analysis

We consider a circuit consisting of two-terminals only. We assume that this circuit is
planar, that is, that you can draw it on a sheet without crossings between components.
Without loss of generality, we assume also that the circuit is not hinged, that is, that it
cannot be divided into two parts connected by a single node. Under these assumptions,
the sides (elements) of the circuit form a set of meshes (i.e., "holes" on the sheet)
bounded by two or more elements of the circuit. If L is the number of elements of
the circuit and N is the number of nodes, it can be shown that the number of meshes
in the circuit is $L - N + 1$. As pointed out in Sect. 3.1, this is also the number of
independent KVL equations for the circuit. This fact is not surprising: for each mesh
you can write a KVL equation in which the voltages are different, in whole or in part,
from those of every other mesh of the circuit. In addition, the KVL equations that can
be written for any other loop of the circuit are linearly dependent on those written
for the meshes. As an example, in the planar circuit of Fig. 7.8 the KVL equations
corresponding to the meshes *I,II,III* are

$$(I) \quad v_a + v_c - v_b = 0$$

$$(II) \quad -v_c + v_d - v_e = 0$$

$$(III) \quad v_e - v_f - v_g = 0$$

The KVL for the loop (not mesh!) formed by the two-terminals c, d, f, g can be
written as $-v_c + v_d - v_f - v_g = 0$, but this equation linearly depends on equations
(II) and *(III)*, because you can get it by summing member by member these equations.
(Note that the sum drops the term v_e.)

The currents through the L two-terminals must meet the constraints given by the
$N - 1$ KCL equations. These currents can be expressed in order to meet the KCLs
automatically. To do this, simply associate a "mesh current"[1] with each mesh and

---

[1]Notice that this is not necessarily a variable that can be physically measured through an amper-
ometer.

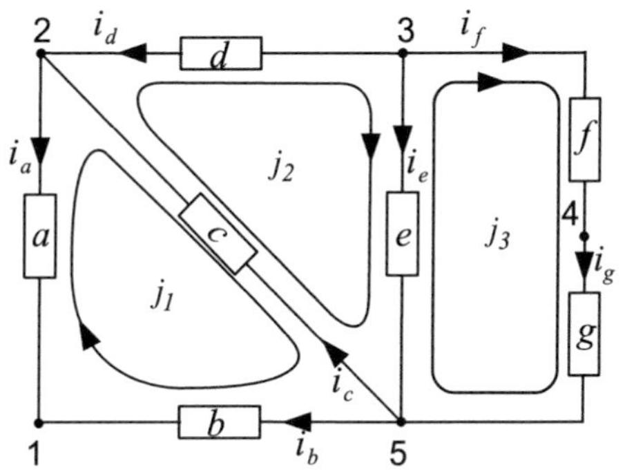

Fig. 7.9 Mesh currents and component currents for the circuit of Fig. 7.8

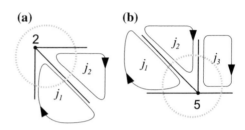

Fig. 7.10 Mesh currents at nodes 2 and 5 for the circuit of Fig. 7.9

(a)　　　　　**(b)**

express the currents through the components in terms of these mesh currents. Then the current in each component will depend on one or two mesh currents.

In the proposed example, you can define the mesh currents j_1, j_2, j_3, which are shown in Fig. 7.9, together with the component currents. In the two-terminals c and e, which are shared by two meshes, the currents are given by the algebraic sum of their mesh currents. With the conventions adopted, the complete set of equations relating the component currents and the mesh currents is as follows.

$$i_a = -j_1; \qquad i_d = -j_2;$$

$$i_b = j_1; \qquad i_e = j_2 - j_3;$$

$$i_c = -j_1 + j_2; \qquad i_f = i_g = j_3.$$

Each mesh current passing through the closed boundary of a cut-set does it twice, one as input and the other as output, for which the resulting contribution to the KCL is zero. (See Fig. 7.10, which represents the cut-sets for nodes 2 and 5 with the pertinent mesh related to the circuit of Fig. 7.9.) Therefore, the mesh currents automatically satisfy the KCL at the nodes (and more generally the KCL for any cut-set in the circuit).

The mesh analysis method considers as unknowns the $(L - N + 1)$ mesh currents instead of the L component currents. After getting the mesh currents, the currents in the components can be easily obtained, as in the example just considered.

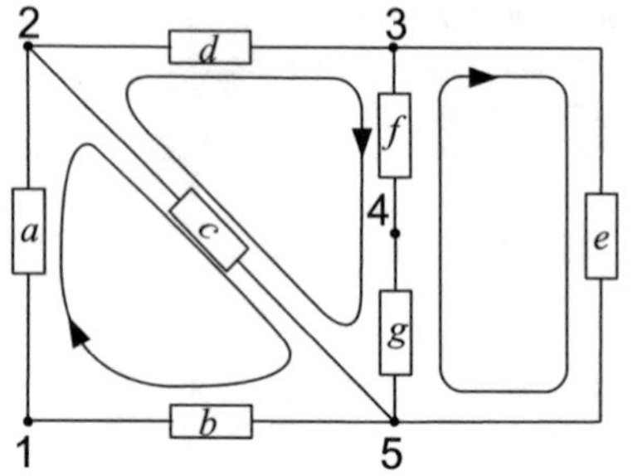

Fig. 7.11 Another way of
drawing the circuit of
Figs. 7.8 and 7.9

Fig. 7.12 Planar circuit for
the Case Study. The mesh
currents j_1, j_2, j_3 are
oriented clockwise

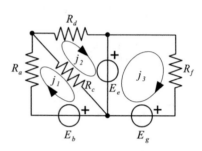

Remark: It should be noted that drawing the same circuit in different ways may originate different meshes. For instance, the circuit of Figs. 7.8 and 7.9 can be redrawn as shown in Fig. 7.11. In this case, the edges f and g form a mesh with edges c, d, and (as before) a mesh with e. The new set of mesh currents is completely equivalent to the previous one.

Assumptions: The circuit is planar and consists only of

• Linear two-terminal components that admit the current basis
• Independent voltage sources

The KVL equations are written with reference to the meshes of the circuit; by means of the descriptive equations of the components, each voltage contribution inside the mesh can be written directly in terms of the mesh currents. The procedure details are illustrated by the following example.

Case Study

Solve with pure mesh analysis the circuit shown in Fig. 7.12.

The circuit is planar and contains linear resistors and independent voltage sources only. The three meshes are associated with the mesh currents j_1, j_2, j_3, respectively, all oriented in the same way. For each mesh, we can write the KVL taking the component voltages as positive if oriented like the mesh current. In this way we obtain the following equations.

$$\begin{cases} -E_b - R_a j_1 - R_c(j_1 - j_2) = 0 \\ -E_e - R_c(j_2 - j_1) - R_d j_2 = 0 \\ E_e - R_f j_3 - E_g = 0. \end{cases}$$

After a few manipulations, the system can be recast as

$$\begin{cases} (R_a + R_c) j_1 - R_c j_2 = -E_b \\ -R_c j_1 + (R_c + R_d) j_2 = -E_e \\ R_f j_3 = E_e - E_g. \end{cases}$$

Therefore, the system of equations can be written in compact form as

$$Rj = \hat{e} \text{ with } j = \begin{pmatrix} j_1 \\ j_2 \\ j_3 \end{pmatrix}; \ \hat{e} = \begin{pmatrix} -E_b \\ -E_e \\ E_e - E_g \end{pmatrix}; \ R = \begin{pmatrix} (R_a + R_c) & -R_c & 0 \\ -R_c & (R_c + R_d) & 0 \\ 0 & 0 & R_f \end{pmatrix}$$

namely, in such a way that the terms on the main diagonal of R appear with a positive sign.

We notice that the circuit equations, *per se*, could be those of a three-port containing only reciprocal components (resistors) and described by the matrix R connected to three equivalent voltage sources, described by the vector $\hat{e}$. (See the equivalent circuit shown in Fig. 7.13.) Due to the reciprocity Theorem 3.1 (see also Sect. 6.3) the three-port is reciprocal. This implies (owing to a generalization of the property shown in Sect. 5.4.1 for two-ports) that the matrix R is symmetric.

Of course, the mesh currents are easily obtained from the system of equations whatever the form in which this system is written. However, the system in the form $Rj = \hat{e}$ is particularly simple to obtain because it rests on few systematic rules.

Fig. 7.13 Equivalent circuit for the Case Study

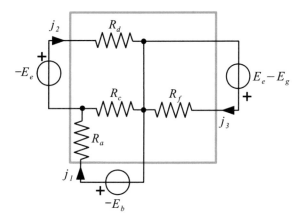

These rules are deduced easily by referring to the considered circuit, but are general in nature and as such they are set out below.

In the system of equations

$$Rj = \hat{e} \tag{7.2}$$

the entries of R and $\hat{e}$ are obtained by circuit inspection according to the rules:

- In the ith row, associated with the ith mesh, the entry R_{ii} of the main diagonal is the sum of the resistances encountered by j_i along the mesh i.
- R_{ik}, with $k \neq i$, is the sum of the resistances (preceded by the $-$ sign) shared by j_i and j_k. If no resistance is shared by j_i and j_k, then $R_{ik} = 0$.
- The ith entry of $\hat{e}$ is the algebraic sum of the impressed voltages in the ith mesh; each of these voltages contributes with the $+$ sign if its direction agrees with that of the mesh current j_i; otherwise it is taken with the $-$ sign.

According to these rules, everything works as if in the mesh associated with j_i only one voltage source were present. The voltage value is given by the ith element of the vector $\hat{e}$: for instance, in the circuit of Fig. 7.12, the equivalent voltage source for the mesh associated with j_3 is $E_e - E_g$.

7.2.2 Modified Mesh Analysis

As for the node analysis, the mesh analysis method can be modified so as to admit the presence of independent current sources and of the four controlled sources. As a general criterion, we need to add proper *auxiliary unknowns* to the *primary unknowns* (the mesh currents $\{j_k\}$).

A first type of auxiliary unknown is related to the current sources, both independent and controlled. Indeed, in both cases the component's equation does not provide information about the voltage across the source. This voltage must therefore be regarded as an unknown v_x (Fig. 7.14), to be found together with the mesh currents.

The second kind of auxiliary unknown is the driving voltage for both VCVS and VCCS. This descriptive variable appears in the component's equation and must be added to the set of variables to be determined. For CCVSs and CCCSs, instead, the driving variable is a current that can be expressed directly in terms of the mesh currents, without the need of a further auxiliary unknown.

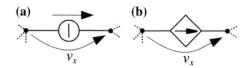

Fig. 7.14 Auxiliary unknown v_x to extend mesh analysis to: **a** independent current source; **b** controlled current source

Fig. 7.15 Planar circuit for
the Case Study

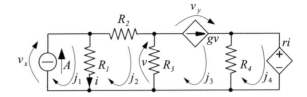

Because of the introduction of the auxiliary unknowns, the system to be solved
does not maintain the simple form Eq. 7.2 as for the case of pure mesh analysis.

Case Study

Solve with the modified mesh analysis the circuit shown in Fig. 7.15.

This planar circuit contains linear resistors, a linear VCCS (driving voltage: v),
a linear CCVS (driving current: i), and an independent current source A. The
circuit equations are expressed in terms of the mesh currents, of the auxiliary
variables v_x and v_y for the two current sources, and of the driving variables i
and v. With this in mind, we can formulate the KVL for each mesh as

$$\begin{cases} v_x - R_1 (j_1 - j_2) = 0 \\ R_1 (j_1 - j_2) - R_2 j_2 - R_3 (j_2 - j_3) = 0 \\ v + v_y - R_4 (j_3 - j_4) = 0 \\ R_4 (j_3 - j_4) - ri = 0 \end{cases}$$

The remaining equations are the driving variables i and v expressed in terms
of mesh currents and the descriptive equations for independent current source
and VCCS:

$$\begin{cases} i = j_1 - j_2 \\ v = R_3 (j_2 - j_3) \\ j_3 = gv \\ j_1 = A \end{cases}$$

Note that the driving current i of the CCVS is expressed in a trivial way in
terms of the mesh currents; furthermore, the position of the current source A
on the external loop of the circuit enables us to assign the value A immediately
to the mesh current j_1.

As $j_1 = A$, the entire system of equations can be reformulated easily as follows. A first group of three equations containing only the unknowns j_2, j_3, j_4:

$$\begin{cases} R_1 A - j_2 (R_1 + R_2 + R_3) + R_3 j_3 = 0 \\ r j_2 + R_4 j_3 - R_4 j_4 - rA = 0 \\ g R_3 j_2 - j_3 (1 + g R_3) = 0 \end{cases}$$

and a second group of equations that, taking j_2, j_3, j_4 as solutions of the previous equations, lead directly to writing the expressions of the remaining unknowns v_x, v_y, i, v:

$$\begin{cases} v_x = R_1 (A - j_2) \\ v_y = -R_3 j_2 + (R_3 + R_4) j_3 - R_4 j_4 \\ i = A - j_2 \\ v = R_3 (j_2 - j_3) \end{cases}$$

7.2.3 Substitution Rule

Writing the mesh equations in a circuit can sometimes become easier by applying the substitution rule described below, which can be used for both independent and controlled sources.

Figure 7.16a represents the portion of a circuit containing a voltage source E. The branch containing E can be replaced by a short circuit (and the nodes 1 and 2 reduced to a single node) by modifying the circuit as shown in Fig. 7.16b, where two voltage sources E are connected in series to the components a and b, respectively. The KVLs for the loops involving $\{E, v_a, v_3\}$ and $\{E, v_b, v_4\}$, respectively, are identical in the original circuit and in its modified version; moreover, the same KCL holds for the cut-set in the original circuit and for the node in the modified one (both indicated by closed gray dotted lines in the figures). Therefore, the transformation does not affect the behavior of the circuit. As a result of this substitution, we have a voltage source connected in series to each of the components a and b. If a is a resistor, the series connection with a voltage source originates a Thévenin equivalent; if a is, in turn, a voltage source, the series connection is equivalent to a single source, whose voltage is given by the algebraic sum of the two constituents. (See Sect. 3.5.2.1.)

As an example, the circuit of Fig. 7.12 can be modified by applying the substitution rule to the voltage source E_e. The resulting circuit is shown in Fig. 7.17.

Fig. 7.16 Removing a branch containing a voltage source: **a** original circuit; **b** equivalent circuit

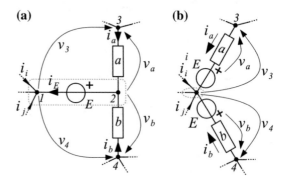

Fig. 7.17 The circuit of
Fig. 7.12 after the application
of the substitution rule to E_e

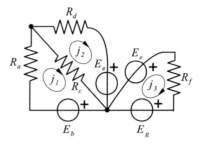

The equivalent voltage sources pertinent to the three current meshes j_1, j_2, j_3 are
$-E_b$, $-E_e$, and $E_e - E_g$. These are also the components of the vector $\hat{e}$ obtained in
Sect. 7.2.1.

7.3 Superposition Principle

The superposition principle is a general property valid for all linear systems. In very
general terms, this principle states that the "response" (output) of a linear system
induced by two or more stimuli (inputs) can be obtained at any instant as the sum of
the responses due separately to each stimulus. In classical mechanics, for instance,
for a point mass initially at rest and subject to the action of two forces (vectors) F_1 and
F_2 – not parallel, in general – the velocity vector at a given time can be obtained by
summing the velocity vectors due to the separate actions of F_1 and F_2. Switching to
another branch of physics, that is, electrostatics, at any point of the empty space and
at any instant the electric potential generated by the action of two electric charges q_1
and q_2 (stimuli) can be obtained by summing the contributions separately originated
by q_1 and q_2.

In other words, if input $x_1(t)$ produces output $y_1(t)$ and input $x_2(t)$ produces
output $y_2(t)$, then input $k_1 x_1(t) + k_2 x_2(t)$ produces output $k_1 y_1(t) + k_2 y_2(t)$, with
k_1 and k_2 dimensionless constants.

Stimuli and responses can be either scalar elements (e.g., numbers, functions) or
vectors. In the latter case, the superposition implies a vector sum.

For linear electric circuits, each stimulus (input) corresponds to an independent
voltage or current source; each term of the response can be either the current through
a branch of the circuit or the voltage across a pair of nodes of the circuit. The
superposition principle for electric circuits can then be expressed as follows.

> **Superposition principle**: The solution (or response, or output) of a circuit
> made up of linear components and two or more (say P) independent sources
> (inputs) is the sum of the outputs that would have been determined by each
> input acting alone, where all the other independent sources are turned off, that
> is, replaced by either a short circuit (for voltage sources, see Fig. 7.18a, b) or

Fig. 7.18 Voltage source (**a**) and its turned-off equivalent (**b**). Current source (**c**) and its turned-off equivalent (**d**)

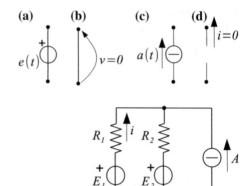

Fig. 7.19 Case Study 1

an open circuit (for current sources, see Fig. 7.18c, d). Then, the solution of the original circuit is the sum of the solutions of P *auxiliary circuits,* each obtained by turning on one independent source in turn and turning off all the other independent sources.

In the most general case, the linear circuit elements can be either memoryless or dynamic.

A more formal expression of this principle (superposition theorem) is provided and demonstrated for time-invariant memoryless circuits in Sect. 8.3.

Case Study 1
Find the current i in the circuit shown in Fig. 7.19.

As stated above, we can solve this problem in two different ways.
Way 1. We apply the superposition principle.

We have three independent sources; thus we have to solve three auxiliary circuits. In the first one, shown in Fig. 7.20a, only the voltage source E_1 is turned on. The corresponding solution (R_1 and R_2 are connected in series in the auxiliary circuit) is $i^{(1)} = \dfrac{E_1}{R_1 + R_2}$.

In the second auxiliary circuit, shown in Fig. 7.20b, only the voltage source E_2 is turned on. The corresponding solution is $i^{(2)} = -\dfrac{E_2}{R_1 + R_2}$.

Finally, in the third auxiliary circuit, shown in Fig. 7.20c, only the current source A is turned on. The corresponding solution (according to the current divider structure of this auxiliary circuit, Sect. 3.6.2) is $i^{(3)} = -A\dfrac{R_2}{R_1 + R_2}$.

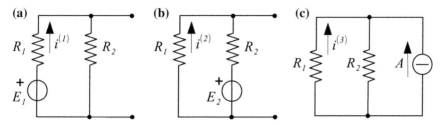

Fig. 7.20 First (a), second (b), and third (c) auxiliary circuits for Case Study 1

Fig. 7.21 Case Study 2

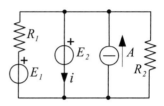

> By applying the superposition principle, the solution of the original circuit
> is $i = i^{(1)} + i^{(2)} + i^{(3)} = \dfrac{E_1 - E_2 - R_2 A}{R_1 + R_2}$.
>
> **Way 2.** We do not apply the superposition principle. You can easily check
> that the solution is the same as above.
> We remark that each of the sources E_1, E_2, A may be constant or time-
> varying.

> **Case Study 2**
> *Apply the superposition principle to find the current i in the circuit of
> Fig. 7.21.*
>
> Consider the auxiliary circuits shown in Fig. 7.22 obtained from the original
> circuit by applying a single source at a time. According to the superposition
> principle, the current i can be determined as the sum of the currents $i^{(1)}$, $i^{(2)}$, $i^{(3)}$
> in the auxiliary circuits.
> Thus we have:
>
> $$i = i^{(1)} + i^{(2)} + i^{(3)} = \frac{E_1}{R_1} - \frac{E_2(R_1 + R_2)}{R_1 R_2} + A.$$
>
> You can check the correctness of this result by writing the circuit equations
> and solving for the current i.

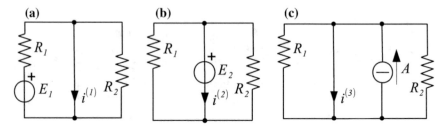

Fig. 7.22 Auxiliary circuits for Case Study 2

Fig. 7.23 Case Study 3

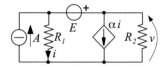

Case Study 3

Apply the superposition principle to find the voltage v in the circuit of Fig. 7.23.

Consider the auxiliary circuits shown in Fig. 7.24, obtained from the original circuit by applying a single independent source at a time. Note that the linear CCCS is present in both circuits. Once more, according to the superposition principle, the voltage v can be determined as the sum of voltages $v^{(1)}$ and $v^{(2)}$ in the auxiliary circuits.

For the auxiliary circuit of Fig. 7.24a we have

$$\begin{cases} A = i(1+\alpha) + \dfrac{v^{(1)}}{R_2} \\ i = \dfrac{v^{(1)}}{R_1} \end{cases} \quad \text{then } v^{(1)} = \frac{A R_1 R_2}{R_1 + R_2(1+\alpha)}.$$

For the auxiliary circuit of Fig. 7.24b we write

$$\begin{cases} v^{(2)} = E + R_1 i \\ i(1+\alpha) + \dfrac{v^{(2)}}{R_2} = 0 \end{cases} \quad \text{then } v^{(2)} = \frac{E R_2(1+\alpha)}{R_1 + R_2(1+\alpha)}.$$

Finally, we have:

$$v = R_2 \frac{A R_1 + E(1+\alpha)}{R_1 + R_2(1+\alpha)}.$$

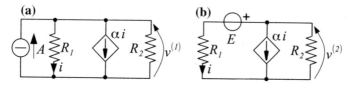

Fig. 7.24 Auxiliary circuits for Case Study 3

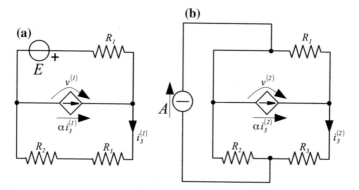

Fig. 7.25 First (**a**) and second (**b**) auxiliary circuits for Case Study 4

Case Study 4

Solve the case study of Sect. 5.2.1.1 by using the superposition principle.

We have two independent sources; thus we have to solve two auxiliary circuits. In the first one, shown in Fig. 7.25a, only the voltage source is turned on. The corresponding solution is $v^{(1)} = E \left(1 + \dfrac{R_1(\alpha - 1)}{R_1(1 - \alpha) + R_2 + R_3} \right)$.

In the second auxiliary circuit, shown in Fig. 7.25b, only the current source is turned on. The corresponding solution is

$$v^{(2)} = R_1(\alpha - 1)\frac{R_2 A}{R_1(1 - \alpha) + R_2 + R_3}.$$

By applying the superposition principle, the solution of the original circuit is

$$v = v^{(1)} + v^{(2)} = E + R_1(\alpha - 1)\frac{E + R_2 A}{R_1(1 - \alpha) + R_2 + R_3}.$$

Actually, we already applied the superposition principle before introducing it: it happened in Sect. 3.4, where we introduced the Thévenin and Norton equivalent representations of two-terminal resistive components. In that case, the superposition principle was applied only partially, by considering a first input external to the two-terminal element (current source i for Thévenin and voltage source v for Norton) and all the internal independent sources as a single second input. You can check your comprehension by solving again the case studies of Sect. 3.4, by also applying the superposition principle to the internal independent sources.

We remark that the superposition principle only works for output voltages and currents **but not for powers**, because the power definition involves a nonlinear operation (product) on voltages and currents. In other words, the sum of the powers absorbed/delivered by a given component due to the action of each single independent source **is not** the real power absorbed/delivered by that component. To calculate this power properly, we should first use the superposition principle to find the current(s) and voltage(s) involved in the power computation and then calculate the power.

Case Study 5

Find the power p delivered by the voltage source E_1 in the circuit shown in Fig. 7.19.

The required power is $p = E_1 i = E_1 \dfrac{E_1 - E_2 - R_2 A}{R_1 + R_2}$. You can easily check that if we applied the superposition principle to the powers delivered in the three auxiliary circuits ($p' = E_1 i^{(1)} + 0 \cdot i^{(2)} + 0 \cdot i^{(3)}$), we would obtain a wrong solution.

7.4   Substitution Principle

Another general principle valid under mild assumptions is the substitution principle. It is yet another tool to substitute with a simpler model a part of a circuit we are not interested to describe in detail. For instance, a housewife is not interested to know what there is behind a wall socket: for her (and for a large part of us!) the socket can be more simply represented by an ideal voltage source, which supplies the energy required by an appliance for working properly.

In the case of memoryless circuits (linear or nonlinear, time-varying or time-invariant), we can prove the following theorem.

Theorem 7.1 *(Substitution theorem) We consider a memoryless circuit (network) $\aleph$ that can be decomposed into two complementary subnetworks (composite two-terminals) S_1 and S_2, as shown in Fig. 7.26. At the parallel connection between S_1 and S_2 we can measure voltage v and current i. We assume that $\aleph$ admits a unique solution (v, i). Then, if S_1 admits the current basis, S_2 can be replaced by an independent current source with impressed current i, as shown in Fig. 7.27a. Analogously, if S_1 admits the voltage basis, then S_2 can be replaced by an independent voltage source with impressed voltage v, as shown in Fig. 7.27b.*

The theorem states that, from the point of view of S_1, there is no difference between the circuit configuration shown in Fig. 7.26 and the one shown in Fig. 7.27 (either a

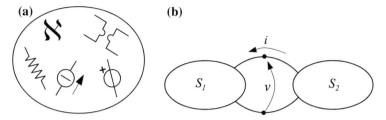

Fig. 7.26 The network $\aleph$ (**a**) and its decomposition into two subnetworks S_1 and S_2 (**b**)

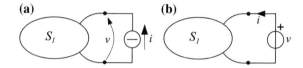

Fig. 7.27 Application of the substitution principle when S_1 admits the current (**a**) or the voltage (**b**) basis

or b). Once more, this means that, if we are not interested in what happens within S_2, we can represent more compactly this part of the original network $\aleph$, thus resorting to a higher-level model.

Proof By assumption, there exists a unique solution (v, i). Then, by definition of basis (Sect. 3.3.4), if the memoryless composite two-terminal S_1 admits the current basis, its descriptive current i can be *arbitrarily* assigned and its descriptive voltage v is obtained *univocally*. Moreover, if the independent current source admits the voltage basis, then it is compatible with any voltage v.

A similar reasoning can be applied, *mutatis mutandis*, if S_1 admits the voltage basis. $\square$

7.5 Practical Rules

In this section we introduce some further practical rules that can be useful to make handling the most common circuit structures simpler.

7.5.1 Millmann's Rule

The reference circuit for the application of Millmann's rule is shown in Fig. 7.28. It consists of a finite number K of branches connected in parallel, each of which contains a Thévenin equivalent, that is, a voltage source E_i in series with a resistor R_i $(i = 1, \dots, K)$.

The voltage v common to all the branches can be easily obtained as follows.

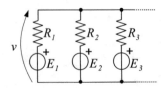

Fig. 7.28 The reference circuit for Millmann's rule

(a)   **(b)**

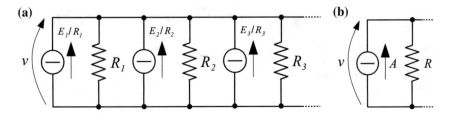

Fig. 7.29 a The circuit of Fig. 7.28 after the replacement of each branch with its Norton equivalent;
b the final equivalent circuit

- Replace each branch with its Norton equivalent, as shown in Fig. 7.29a. The current impressed by the ith Norton current source is E_i/R_i.
- Determine the resulting current source A and the equivalent resistance R (Fig. 7.29b); the voltage v equals AR.

The expressions for A and R are

$$A = \frac{E_1}{R_1} + \frac{E_2}{R_2} + \frac{E_3}{R_3} + \cdots ; \qquad R = \frac{1}{\dfrac{1}{R_1} + \dfrac{1}{R_2} + \dfrac{1}{R_3} + \cdots}.$$

Therefore, the resulting expression of the voltage v is

$$v = \frac{\dfrac{E_1}{R_1} + \dfrac{E_2}{R_2} + \dfrac{E_3}{R_3} + \cdots}{\dfrac{1}{R_1} + \dfrac{1}{R_2} + \dfrac{1}{R_3} + \cdots}. \qquad (7.3)$$

This result can also be formulated to emphasize the contributions of the individual voltage sources as they come out from the superposition principle:

$$v = \alpha_1 E_1 + \alpha_2 E_2 + \alpha_3 E_3 + \cdots \quad \text{with } \alpha_i = \frac{1}{R_i} \frac{1}{\dfrac{1}{R_1} + \dfrac{1}{R_2} + \dfrac{1}{R_3} + \cdots} \quad (i = 1, \cdots, K).$$

Fig. 7.30 Example of generalization of Millmann's rule

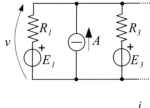

Fig. 7.31 T structure

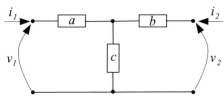

7.5.1.1 Some Comments and Generalizations

Once the voltage v is known, the current through each resistor R_i can be obtained immediately. This is evident observing the circuit of Fig. 7.29a.

If one of the branches (say the kth branch) in the circuit of Fig. 7.28 contains only the resistor R_k (i.e., when $E_k = 0$), we must remove the term E_k/R_k at the numerator of Eq. 7.3.

Millmann's rule can be easily extended to the case in which one or more branches of the circuit are constituted by a current source. For example, in the circuit of Fig. 7.30 the expression Eq. 7.3 changes as follows.

$$v = \frac{\dfrac{E_1}{R_1} + A + \dfrac{E_3}{R_3}}{\dfrac{1}{R_1} + \dfrac{1}{R_3}}$$

7.5.2 Two-Ports with $T - \Pi$ Structure

Among the most common two-ports containing two-terminals as building blocks, the ones shown in Figs. 7.31 and 7.32 are particularly interesting. The first one is usually known as a T (or "star", or "wye", or Y) structure; the second one is known as a Π (or "mesh", or "delta", or Δ) structure. In both cases, the ports have a common terminal. In the following, unless otherwise specified, we adopt the term T for the first and the term Π for the second structure. In the general case, the two-terminal elements a, b, c may be of a different kind, but here we focus on the case where they are linear resistors.

Figure 7.33 shows a T structure with three resistors r_a, r_b, r_c. The resistance matrix R of the resulting two-port is defined as

Fig. 7.32 Π structure

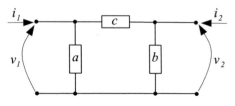

Fig. 7.33 Resistive T
structure

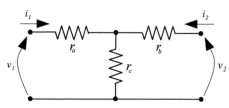

$$\begin{pmatrix} v_1 \\ v_2 \end{pmatrix} = R \begin{pmatrix} i_1 \\ i_2 \end{pmatrix} = \begin{pmatrix} R_{11} & R_{12} \\ R_{21} & R_{22} \end{pmatrix} \begin{pmatrix} i_1 \\ i_2 \end{pmatrix},$$

where

$$R = \begin{pmatrix} r_a + r_c & r_c \\ r_c & r_b + r_c \end{pmatrix} \tag{7.4}$$

Note that the two-port is reciprocal ($R_{12} = R_{21}$). This is a direct consequence of
the fact that all components are reciprocal. (See Theorem 3.1.)

From Eq. 7.4 we have:

$$r_a = R_{11} - R_{12}; \quad r_b = R_{22} - R_{12}; \quad r_c = R_{12} = R_{21}. \tag{7.5}$$

Therefore, given the matrix R of a reciprocal two-port and assuming that the ports
have a common terminal, you can get the resistance values (r_a, r_b, r_c) of the elements
of the equivalent T structure. This structure can be realized by standard resistors as
long as the resistance values are positive (or null). Therefore, the following inequalities must be satisfied:

$$R_{12} \geq 0; \quad R_{11} \geq R_{12}; \quad R_{22} \geq R_{12}.$$

When $r_a = r_b = 0$ and $r_c > 0$, the T structure becomes that of Fig. 7.34 and the
corresponding matrix R is

$$R = r_c \begin{pmatrix} 1 & 1 \\ 1 & 1 \end{pmatrix},$$

as you can easily verify.

Figure 7.35 shows a Π structure with three resistors $1/g_a, 1/g_b, 1/g_c$. The conductance matrix G of the resulting two-port is defined as

Fig. 7.34 Resistive T structure: a particular case

Fig. 7.35 Resistive Π structure

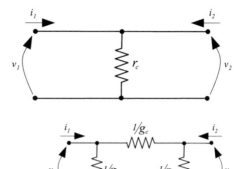

$$\begin{pmatrix} i_1 \\ i_2 \end{pmatrix} = G \begin{pmatrix} v_1 \\ v_2 \end{pmatrix} = \begin{pmatrix} G_{11} & G_{12} \\ G_{21} & G_{22} \end{pmatrix} \begin{pmatrix} v_1 \\ v_2 \end{pmatrix},$$

where

$$G = \begin{pmatrix} g_a + g_c & -g_c \\ -g_c & g_b + g_c \end{pmatrix} \tag{7.6}$$

Note that the two-port is reciprocal ($G_{12} = G_{21}$). This is because all components are reciprocal. (See Theorem 3.1.)

From Eq. 7.6 we have:

$$g_a = G_{11} + G_{12}; \quad g_b = G_{22} + G_{12}; \quad g_c = -G_{12} = -G_{21}. \tag{7.7}$$

Therefore, given the matrix G of a reciprocal two-port and assuming that the ports have a common terminal, you can get the conductance values (g_a, g_b, g_c) of the elements of the equivalent Π structure. This structure can be realized by standard resistors as long as the conductance values are positive (or null). Therefore, the following inequalities must be satisfied.

$$G_{12} \le 0; \quad G_{11} \ge -G_{12}; \quad G_{22} \ge -G_{12}.$$

When $g_a = g_b = 0$ and $g_c > 0$, the Π structure becomes that of Fig. 7.36 and the corresponding matrix G is

$$G = g_c \begin{pmatrix} 1 & -1 \\ -1 & 1 \end{pmatrix},$$

as you can easily verify.

Fig. 7.36 Resistive Π
structure: a particular case

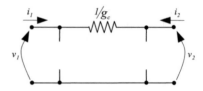

7.5.3 $T \rightleftarrows \Pi$ Transformations

Suppose that a given two-port can be represented equivalently by both a T-structure (R matrix) and a Π-structure (G matrix). In this case, we must have $GR = I$, where I denotes the 2×2 identity matrix. This is the basis for expressing the parameters of the Π structure in terms of those of the T structure and vice versa.

7.5.3.1 $T \rightarrow \Pi$ Transformation

In this case we consider as known the parameters r_a, r_b, r_c of the T structure of Fig. 7.33. To obtain the parameters g_a, g_b, g_c of the corresponding Π structure (Fig. 7.35), we first set $G = R^{-1}$. From Eqs. 7.4 and 7.6 we have

$$\begin{pmatrix} g_a + g_c & -g_c \\ -g_c & g_b + g_c \end{pmatrix} = \begin{pmatrix} r_a + r_c & r_c \\ r_c & r_b + r_c \end{pmatrix}^{-1} = \frac{1}{det(R)} \begin{pmatrix} r_b + r_c & -r_c \\ -r_c & r_a + r_c \end{pmatrix}$$

where $det(R) = r_a r_b + r_a r_c + r_b r_c$ is the determinant of R. Now, taking into account Eq. 7.7, it is easy to obtain g_a, g_b, g_c. For example, we have

$$g_a = G_{11} + G_{12} = \frac{r_b}{det(R)} = \frac{r_b}{r_a r_b + r_a r_c + r_b r_c}$$

and the corresponding resistance value is $\dfrac{1}{g_a} = r_a + r_c + \dfrac{r_a r_c}{r_b}$. The other terms are obtained similarly. The complete set of equivalences is shown in Fig. 7.37.

7.5.3.2 $\Pi \rightarrow T$ Transformation

Taking as known the parameters g_a, g_b, g_c of the Π structure, we set $R = G^{-1}$ using Eqs. 7.4 and 7.6:

$$\begin{pmatrix} r_a + r_c & r_c \\ r_c & r_b + r_c \end{pmatrix} = \begin{pmatrix} g_a + g_c & -g_c \\ -g_c & g_b + g_c \end{pmatrix}^{-1} = \frac{1}{det(G)} \begin{pmatrix} g_b + g_c & g_c \\ g_c & g_a + g_c \end{pmatrix}$$

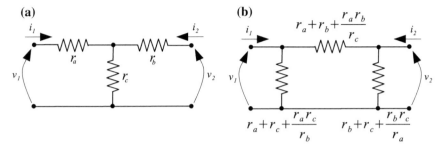

Fig. 7.37 $T \rightarrow \Pi$ transformation. Parameters of the T structure (a); parameters of the Π structure (b)

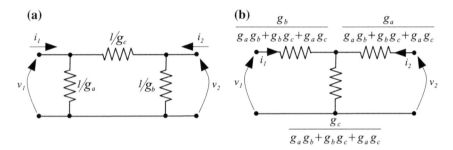

Fig. 7.38 $\Pi \rightarrow T$ transformation. Parameters of the Π structure (a); parameters of the T structure (b)

where $det\,(G) = g_a g_b + g_b g_c + g_a g_c$ is the determinant of G. Now, taking into account Eq. 7.5, we easily obtain r_a, r_b, r_c. For example:

$$r_a = R_{11} - R_{12} = \frac{g_b}{det\,(G)} = \frac{g_b}{g_a g_b + g_b g_c + g_a g_c} = \frac{1}{g_a + g_b + \dfrac{g_a g_c}{g_b}}$$

The other terms are obtained similarly. The complete set of equivalences is shown in Fig. 7.38.

7.5.4 Lattice (Bridge) Structures

The structure shown in Fig. 7.39a is called a lattice structure for obvious reasons. It can be redrawn as in Fig. 7.39b, that is, in the form of the so-called bridge structure. One or the other representation is adopted, depending on convenience.

In order to find the elements of the matrix R, it is appropriate to refer to the bridge representation of Fig. 7.39b. The resistance matrix R of the two-port is written as

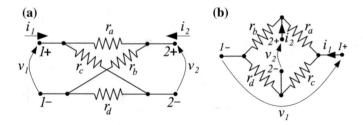

Fig. 7.39 a Lattice structure; **b** equivalent bridge representation

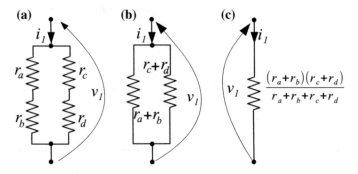

Fig. 7.40 Steps for obtaining R_{11}

$$\begin{pmatrix} v_1 \\ v_2 \end{pmatrix} = R \begin{pmatrix} i_1 \\ i_2 \end{pmatrix} = \begin{pmatrix} R_{11} & R_{12} \\ R_{21} & R_{22} \end{pmatrix} \begin{pmatrix} i_1 \\ i_2 \end{pmatrix}.$$

For the first element R_{11}, we have

$$R_{11} = \frac{v_1}{i_1}\bigg|_{i_2=0},$$

that is, the overall resistance of the structure represented in Fig. 7.40a. The resistance can be obtained according to the steps represented in Fig. 7.40b, c. Finally, we have:

$$R_{11} = \frac{(r_a + r_b)(r_c + r_d)}{r_a + r_b + r_c + r_d}.$$

For R_{12}, we have

$$R_{12} = \frac{v_1}{i_2}\bigg|_{i_1=0}$$

Then, making reference to the circuit structure of Fig. 7.41, we easily obtain:

$$v_1 = r_b i_b - r_a i_a = \frac{i_2}{r_a + r_b + r_c + r_d}(r_b r_c - r_a r_d); \quad R_{12} = \frac{r_b r_c - r_a r_d}{r_a + r_b + r_c + r_d}.$$

Fig. 7.41 Auxiliary two-port analyzed to obtain R_{12}

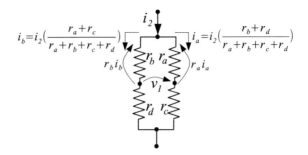

Fig. 7.42 Equivalent representation of a balanced bridge

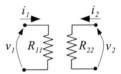

Inasmuch as all the elements of the bridge are two-terminal resistors, the structure is reciprocal (Theorem 3.1) and we have $R_{21} = R_{12}$. With a procedure similar to that followed for R_{11}, we finally have

$$R_{22} = \frac{(r_a + r_c)(r_b + r_d)}{r_a + r_b + r_c + r_d}.$$

As a general comment, the minus sign in the numerator of the expression of R_{12} and R_{21} implies the possibility that these terms are null or negative, in spite of the positiveness of r_a, r_b, r_c, r_d.

When the condition $r_a r_d = r_b r_c$ holds, we have $R_{12} = R_{21} = 0$ and the bridge is called *balanced*. In this case, the matrix R becomes:

$$R = \begin{pmatrix} R_{11} & 0 \\ 0 & R_{22} \end{pmatrix} \quad \text{with} \quad R_{11} = \frac{r_c(r_a + r_b)}{r_a + r_c} \quad \text{and} \quad R_{22} = \frac{r_b(r_a + r_c)}{r_a + r_b}.$$

In this case, the two ports become uncoupled (i.e., the two-port is zero-directional) and the bridge reduces to two separate resistors, as shown in Fig. 7.42.

Another situation of interest is when one port (say port 1) is connected to a voltage source E_1 and the second port is left unconnected (open-ended). This is shown in Fig. 7.43. Under this condition, the voltage ratio α between v_2 and v_1 is

$$\alpha = \left. \frac{v_2}{v_1} \right|_{i_2=0} = \left. \frac{v_c - v_a}{E_1} \right|_{i_2=0} = \frac{r_b r_c - r_a r_d}{(r_a + r_b)(r_c + r_d)}.$$

When the balance condition $r_a r_d = r_b r_c$ holds, we obviously have $\alpha = 0$; that is, $v_2 = 0$ for any v_1.

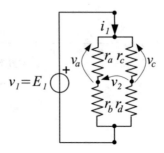

Fig. 7.43 Open-ended bridge

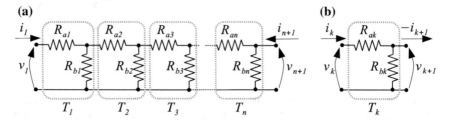

Fig. 7.44 a Ladder structure; **b** a single cell of the ladder

7.5.5 *(Resistive) Ladder Structures*

The resistive structure shown in Fig. 7.44a is very appropriately denoted as a ladder structure. It can be thought of as a chain of a generic number of two-ports called cells (in general different from each other for parameters). The kth cell is represented in Fig. 7.44b together with its port variables.

The voltages and currents in a ladder network within a circuit can be easily studied by choosing one among many methods. One of them is based on the transmission matrices $T_1, \ldots, T_n$ of the cells. Referring to Fig. 7.44b and to Sect. 5.3.4 the matrix T_k for the kth cell is:

$$T_k = \begin{pmatrix} 1 + \dfrac{R_{ak}}{R_{bk}} & R_{ak} \\[2mm] \dfrac{1}{R_{bk}} & 1 \end{pmatrix} \tag{7.8}$$

The way to use the matrices T_k ($k = 1, \ldots, n$) is described in the case study below. For the sake of comparison, the analysis of the same circuit is also carried out with another method.

Case Study

In the circuit of Fig. 7.45, find the voltages v_3 and v_4 in terms of the source voltage E. The resistance values of the ladder are: $R_{a1} = 5\Omega$; $R_{a2} = R_{a3} = 3\Omega$; $R_{b1} = R_{b2} = 2\Omega$; $R_{b3} = 1\Omega$.

Way 1. For the assigned parameter values, from Eq. 7.8 we obtain the matrices T_1, T_2, T_3:

$$T_1 = \begin{pmatrix} \frac{7}{2} & 5 \\ \frac{1}{2} & 1 \end{pmatrix}; \quad T_2 = \begin{pmatrix} \frac{5}{2} & 3 \\ \frac{1}{2} & 1 \end{pmatrix}; \quad T_3 = \begin{pmatrix} 4 & 3 \\ 1 & 1 \end{pmatrix}.$$

Then, the overall transmission matrix T of the ladder is:

$$T = T_1 T_2 T_3 = \begin{pmatrix} \frac{121}{2} & \frac{197}{4} \\ \frac{19}{2} & \frac{31}{4} \end{pmatrix}.$$

Because $v_1 = E$ and $i_4 = 0$, we immediately write $E = T_{11}v_4 + T_{12} \cdot 0$; then $v_4 = \dfrac{2}{121} E$. By this result and making use of T_3 we write $v_3 = T_{3_{11}} v_4 + T_{3_{12}} \cdot 0$ and finally $v_3 = 4v_4 = \dfrac{8}{121} E$. The values of the other voltages and currents in the circuit can be found similarly.

Way 2. All the voltages and currents of the circuit are expressed preliminarily in terms of the unknown voltage v_4 at the end of the ladder, going up with an iterative procedure from v_4 to the voltage E. The various steps are easily described in sequence making reference to the ladder elements in Fig. 7.45 and observing that R_{b3} and R_{a3} are connected in series:

$$i_3 = \frac{v_4}{R_{b3}}; \quad\quad v_3 = (R_{a3} + R_{b3})i_3;$$

$$i_2 = i_3 + \frac{v_3}{R_{b2}}; \quad v_2 = R_{a2}i_2 + v_3;$$

$$i_1 = i_2 + \frac{v_2}{R_{b1}}; \quad E = R_{a1}i_1 + v_2.$$

Using the numerical values of the parameters, the previous expressions give:

$$v_3 = 4v_4; \quad v_2 = 13v_4; \quad E = \frac{121}{2}v_4.$$

Fig. 7.45 Case Study

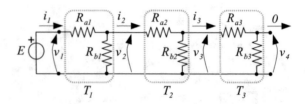

Fig. 7.46 Two equivalent definitions for iterative resistance

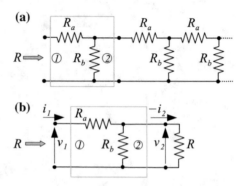

The last equation implies $v_4 = \dfrac{2}{121} E$; therefore $v_3 = 4v_4 = \dfrac{8}{121} E$. The other voltages and currents follow similarly.

Way 3. A third method to solve this circuit makes use of three mesh currents. The details of the solution are left to the reader.

7.5.5.1 Iterative Resistance

Iterative resistance is the resistance R at one port of a two-port when the other port is connected to an infinite chain of identical two-ports, as shown in Fig. 7.46a. Because the chain is assumed to be infinite, iterative resistance is also the resistance value R of an equivalent resistor connected to port 2, as shown in Fig. 7.46b.

Making reference to Fig. 7.46b, the value R of the iterative resistance must fulfill the condition:

$$R = R_a + \frac{R R_b}{R + R_b} \quad \text{that is:} \quad R^2 - R R_a - R_a R_b = 0.$$

The two solutions of this second-degree algebraic equation are:

$$\frac{R_a}{2} \left(1 \pm \sqrt{1 + 4\frac{R_b}{R_a}} \right).$$

Fig. 7.47 Iterative
resistance for port 2

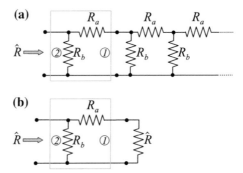

R must be positive, therefore the solution to choose is that with the plus sign:

$$R = \frac{R_a}{2}\left(1 + \sqrt{1 + 4\frac{R_b}{R_a}}\right).$$

In general, the iterative resistance of port 1 is not equal to the iterative resistance of port 2: an infinite chain of two-ports beginning with port 2, as shown in Fig. 7.47a, has an iterative resistance $\hat{R}$ that can be calculated through the circuit of Fig. 7.47b. The obvious condition is:

$$\hat{R} = \frac{R_b(R_a + \hat{R})}{R_b + R_a + \hat{R}} \quad \text{that is} \quad \hat{R}^2 + \hat{R}R_a - R_b R_a = 0.$$

The only positive solution for this equation is:

$$\hat{R} = \frac{R_a}{2}\left(-1 + \sqrt{1 + 4\frac{R_b}{R_a}}\right).$$

Therefore we have $\hat{R} \neq R$ (more precisely, we have $\hat{R} < R$ as you can immediately check).

As a final remark, we observe that the cells of Figs. 7.46a and 7.47a are only two of the possible building blocks for an infinite ladder. The general constituent two-port has the "T" structure delimited by the gray box in Fig. 7.48, where δ is a parameter that ranges in the domain [0, 1]. The previously considered cells, in particular, correspond to the parameter values $\delta = 1$ and $\delta = 0$, respectively. For any value of δ within the range [0, 1], the condition giving the iterative resistance R is

$$R = R_a\delta + \frac{R_b\left[(1 - \delta)R_a + R\right]}{(1 - \delta)R_a + R + R_b}.$$

As an exercise, you can find the corresponding second-degree algebraic equation and check that its positive solutions for $\delta = 1$ and $\delta = 0$ correspond to the previously found values.

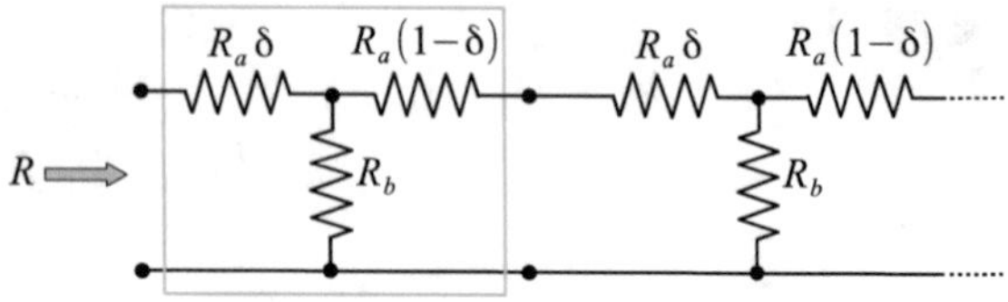

Fig. 7.48 The general
structure of the building
block (*gray box*) for an
infinite ladder

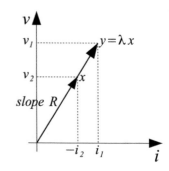

Fig. 7.49 Geometrical
interpretation of the
condition Eq. 7.9 in the (i, v)
plane

7.5.5.2 Iterative Resistance and the Cell's T Matrix

The iterative resistance of the cell is closely related to the eigenvalues of the transmission matrix T. For the cell represented in Fig. 7.46b, the physical condition defining the iterative value R is:

$$\frac{v_1}{i_1} = \frac{v_2}{-i_2} = R \tag{7.9}$$

then the two vectors $y = (v_1 \ i_1)^T$ and $x = (v_2 \ -i_2)^T$ must be *collinear*; that is, $y = \lambda x$, as sketched in Fig. 7.49. Their proportionality can now be interpreted in terms of the general relation $y = Tx$:

$$\begin{cases} y = \lambda x \\ y = Tx \end{cases} ; \quad \text{then} \ \ Tx = \lambda x.$$

Therefore the voltage and current components of both x and y must be thought of as the components of an eigenvector of the matrix T. In terms of the vector x, we have

$$\begin{pmatrix} 1 + \dfrac{R_a}{R_b} & R_a \\ \dfrac{1}{R_b} & 1 \end{pmatrix} \begin{pmatrix} v_2 \\ -i_2 \end{pmatrix} = \lambda \begin{pmatrix} v_2 \\ -i_2 \end{pmatrix} \tag{7.10}$$

where the T elements follow directly from Eq. 7.8.

The characteristic equation of T is

$$\det(\lambda I - T) = \lambda^2 R_b - \lambda(2R_b + R_a) + R_b = 0$$

that is, a second-order polynomial equation. Its solutions are the eigenvalues

$$\lambda = 1 + \frac{1}{2}\frac{R_a}{R_b}\left(1 \pm \sqrt{1 + 4\frac{R_b}{R_a}}\right). \tag{7.11}$$

For both eigenvalues the two scalar Eq. 7.10 become linearly dependent. The second equation (the simplest), in particular, is:

$$\frac{v_2}{R_b} + (-i_2) = \lambda(-i_2)$$

which implies that

$$\frac{v_2}{-i_2} = (\lambda - 1)R_b$$

According to Eq. 7.9, the l.h.s. of this expression is the iterative resistance R, which is positive if and only if $\lambda > 1$, that is, when we choose the solution with the plus sign in Eq. 7.11. The resulting expression for R is

$$R = \frac{R_a}{2}\left(1 + \sqrt{1 + 4\frac{R_b}{R_a}}\right),$$

as expected.

7.6 Problems

7.1 Find the expression and the value of the current i for the circuit shown in Fig. 7.50a, where $R_1 = R_3 = 1k\Omega$, $R_2 = 3k\Omega$, $E_1 = 1V$, $E_2 = 4V$, by applying (you have to analyze the same circuit twice):

• The pure mesh analysis
• The superposition principle

Compare the solution procedures with that used to solve Problem 3.6.

7.2 For the circuit shown in Fig. 7.50b, find:

1. i
2. v
3. Power absorbed by the current source
4. Power absorbed by the voltage source

by applying (you have to analyze the same circuit three times):

• The modified nodal analysis

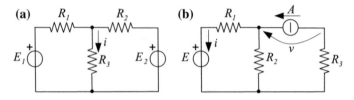

Fig. 7.50 Problems 7.1 **(a)** and 7.2 **(b)**

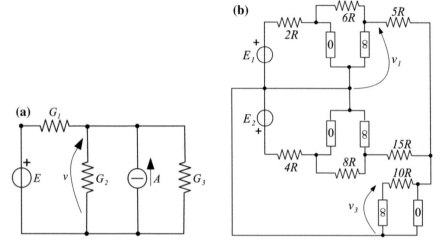

Fig. 7.51 Problems 7.3 **(a)** and 7.4 **(b)**

- The modified mesh analysis
- The superposition principle

Compare the solution procedures with that used to solve Problem 3.7.

7.3 Find the voltage v in the circuit shown in Fig. 7.51a, by applying (you have to analyze the same circuit three times):

- The modified nodal analysis
- The modified mesh analysis
- The superposition principle

Also find the numerical solution for $G_1 = 6\,\mathrm{m\Omega}^{-1}, G_2 = 13\,\mathrm{m\Omega}^{-1}, G_3 = 10\,\mathrm{m\Omega}^{-1}$ $(R_i = 1/G_i, i = 1, 2, 3)$, $E = 24$ V, $A = 1$ mA. Compare the solution procedures with that used to solve Problem 3.12.

7.4 For the circuit shown in Fig. 7.51b, find:

1. The power delivered by voltage source E_1
2. The voltage v_1
3. The voltage v_3

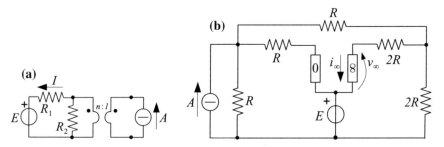

Fig. 7.52 Problems 7.5 (**a**) and 7.6 (**b**)

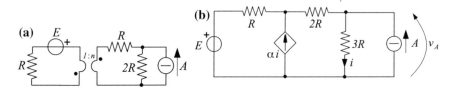

Fig. 7.53 Problems 7.7 (**a**) and 7.8 (**b**)

by applying the superposition principle. Also find the numerical solutions for $E_1 = 2$ V, $R = 1$ kΩ, $E_2 = 3$ V. Compare the solution procedure with that used to solve Problem 5.6.

7.5 For the circuit shown in Fig. 7.52a, find the current I by applying the superposition principle. Also find the numerical solution for $E = 1$ V, $R_1 = 100\Omega$, $R_2 = 400\Omega$, $A = 20$ mA, $n = 2$. You can also solve the same problem by applying the modified nodal analysis and the modified mesh analysis, after substituting the circuit part at the right of R_2 with an equivalent independent current source. Compare the solution procedure with that used to solve Problem 5.7.

7.6 For the circuit shown in Fig. 7.52b, find:

1. The current i_∞
2. The voltage v_∞
3. The power delivered by the voltage source
4. The power absorbed by the current source

by applying the superposition principle. Compare the solution procedure with that used to solve Problem 5.12.

7.7 For the circuit shown in Fig. 7.53a, find, by applying the superposition principle:

1. The power delivered by the voltage source
2. The power absorbed by the ideal transformer

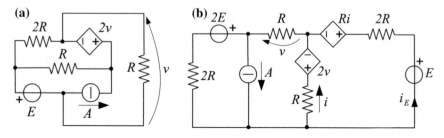

Fig. 7.54 Problems 7.9 (**a**) and 7.10 (**b**)

Also find the numerical solutions for $E = 1$ V, $R = 50\Omega$, $A = 1$ mA, $n = 2$. Compare the solution procedure with that used to solve Problem 5.13.

7.8 Find the voltage v_A in the circuit shown in Fig. 7.53b, by applying (you have to analyze the same circuit three times):

1. The modified nodal analysis
2. The modified mesh analysis
3. The superposition principle

Compare the solution procedures with that used to solve Problem 5.18.

7.9 For the circuit shown in Fig. 7.54a, find:

1. Voltage v
2. Power delivered by the VCVS

by applying (you have to analyze the same circuit three times):

- The modified nodal analysis
- The modified mesh analysis
- The superposition principle

Also find the numerical solutions for $E = 0.6$ V, $R = 30\Omega$, $A = 30$ mA. Compare the solution procedures with that used to solve Problem 5.20.

7.10 For the circuit shown in Fig. 7.54b, find:

1. The current i_E
2. The power absorbed by the current source

by applying (you have to analyze the same circuit three times):

- The modified nodal analysis
- The modified mesh analysis
- The superposition principle

Also find the numerical solutions for $E = 2$ V, $R = 50\Omega$, $A = 300m A$. Compare the solution procedures with that used to solve Problem 5.22.

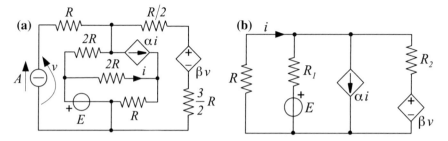

Fig. 7.55 Problems 7.11 (**a**) and 7.12 (**b**)

Fig. 7.56 Problem 7.13

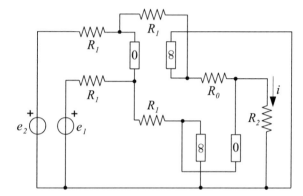

7.11 Find the power absorbed by the CCCS in the circuit shown in Fig. 7.55a by applying (you have to analyze the same circuit three times):

- The modified nodal analysis
- The modified mesh analysis
- The superposition principle

Compare the solution procedures with that used to solve Problem 5.25.

7.12 For the circuit shown in Fig. 7.55b, find the power absorbed by the voltage source (for $\beta = 1$, $\alpha = 0$, $R_1 = R$), by applying (you have to analyze the same circuit three times):

- The modified nodal analysis
- The modified mesh analysis
- The superposition principle

Compare the solution procedures with that used to solve Problem 5.27.

7.13 For the circuit shown in Fig. 7.56, find the current i by applying the superposition principle. Compare the solution procedure with that used to solve Problem 5.28.

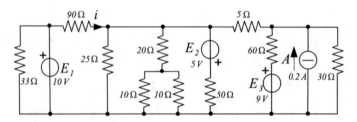

Fig. 7.57 Problem 7.14

Fig. 7.58 Problem 7.15

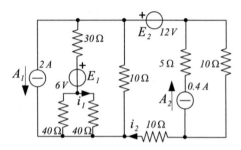

Fig. 7.59 Problem 7.16

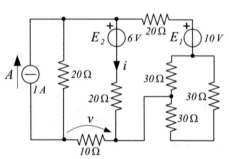

7.14 For the circuit shown in Fig. 7.57, with $E_1 = 10$ V, $E_2 = 5$ V, $E_3 = 9V$, $A = 200$ mA, find the numerical values of:

1. The current i
2. The power delivered by the voltage source E_1

7.15 For the circuit shown in Fig. 7.58, with $A_1 = 2$ A, $A_2 = 400$ mA, $E_1 = 6$ V, $E_2 = 12$ V, find the numerical values of:

1. The current i_1
2. The current i_2
3. The power delivered by the voltage source E_1

7.16 For the circuit shown in Fig. 7.59, with $A = 1$ A, $E_1 = 10$ V, $E_2 = 6$ V, find the numerical values of:

Fig. 7.60 Problem 7.17

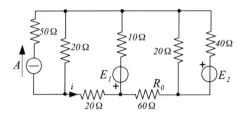

1. The power delivered by the voltage source E_1
2. The current i
3. The voltage v

7.17 For the circuit shown in Fig. 7.60, with $A = 500$ mA, $E_1 = 10$ V, $E_2 = 5$ V, find the numerical values of:

1. The current i
2. The power absorbed by the resistor R_0

Chapter 8
Advanced Concepts

Abstract In this chapter, we shall propose a generalization of the mesh method, describe the tableau method and provide a proof of the superposition principle. Moreover, we shall introduce the Thévenin and Norton equivalent representations of N-port memoryless components.

8.1 From Mesh to Loop Currents: A Graph-Based Generalization

One of the necessary conditions to carry out the mesh analysis of a circuit (consisting only of two-terminal elements) is that it can be drawn on a sheet without crossings between components. This means that its graph is planar and allows unambiguously identifying the meshes (which are independent and form a basis) and drawing all the mesh currents.

A more general formulation, also valid for nonplanar circuits (graphs), can now be given. According to this formulation, the mesh currents can be viewed as a particular case of the so-called loop currents. Each loop current flows through a specific loop in the circuit's graph, and its orientation (in general, free) can be fixed in an unambiguous way even when the graph is not planar.

The structure of each loop associated with a loop current depends on the initial partition into a tree and its respective cotree chosen for the graph. For a directed graph with L edges and N nodes, the number of loop currents is $L - N + 1$. This number coincides with the dimension of the null space $\mathcal{N}(A)$ of the matrix A. (See Sect. 2.2.4.) The set of loop currents constitutes one of the possible bases for $\mathcal{N}(A)$.

This can be easily shown through an example. Consider the graph of Fig. 8.1a. For the tree choice shown in Fig. 8.1b, following the procedure defined in Sect. 2.2.1, the cut-set matrix A is

© Springer International Publishing AG 2018   227
M. Parodi and M. Storace, *Linear and Nonlinear Circuits:*
Basic & Advanced Concepts, Lecture Notes in Electrical Engineering 441,
DOI 10.1007/978-3-319-61234-8_8

Fig. 8.1 A directed graph
(**a**) and its chosen partition
into a tree (*thick grey edges*)
and cotree (*black edges*) (**b**)

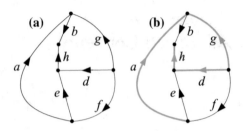

$$A = \begin{array}{c} \\ a \\ g \\ d \\ h \end{array}\begin{array}{ccc|cccc} b & e & f & a & g & d & h \\ 0 & 1 & -1 & 1 & 0 & 0 & 0 \\ -1 & -1 & 1 & 0 & 1 & 0 & 0 \\ 1 & 1 & 0 & 0 & 0 & 1 & 0 \\ 1 & 0 & 0 & 0 & 0 & 0 & 1 \end{array} = (\alpha | I_{N-1}) \qquad (8.1)$$

and we have

$$i = \begin{pmatrix} i_C \\ i_T \end{pmatrix}; \quad i_C = \begin{pmatrix} i_b \\ i_e \\ i_f \end{pmatrix}; \quad i_T = \begin{pmatrix} i_a \\ i_g \\ i_d \\ i_h \end{pmatrix}; \quad Ai = A \begin{pmatrix} i_C \\ i_T \end{pmatrix} = \alpha i_C + i_T = \begin{pmatrix} 0 \\ 0 \\ 0 \\ 0 \end{pmatrix}.$$

Therefore, each current vector i compatible with the graph must be such that $i_T = -\alpha i_C$. (See Sect. 2.2.4.) From Eq. 8.1, the submatrix α is

$$\alpha = \begin{pmatrix} 0 & 1 & -1 \\ -1 & -1 & 1 \\ 1 & 1 & 0 \\ 1 & 0 & 0 \end{pmatrix} = (x_1\ x_2\ x_3) \text{ with } x_1 = \begin{pmatrix} 0 \\ -1 \\ 1 \\ 1 \end{pmatrix}; \quad x_2 = \begin{pmatrix} 1 \\ -1 \\ 1 \\ 0 \end{pmatrix}; \quad x_3 = \begin{pmatrix} -1 \\ 1 \\ 0 \\ 0 \end{pmatrix}.$$

The column vectors x_1, x_2, x_3 can be used to obtain the compatible vector i_T of currents on the tree edges in terms of the i_C components. For any given set i_C of currents on the cotree edges b, e, f we have:

$$i_T = -\alpha i_C = -i_b x_1 - i_e x_2 - i_f x_3.$$

Now, by setting to zero in the previous expression all the coefficients i_b, i_e, i_f except one in all possible ways, we can obtain each loop current and the structure of the loop where it flows in the graph.

- For $i_b = j_1 \neq 0$ and $i_e = i_f = 0$, we have $i_C = j_1 (1\ 0\ 0)^T$ and

$$i_T = -j_1 x_1 = j_1 (0\ 1\ -1\ -1)^T; \quad i = \begin{pmatrix} i_C \\ i_T \end{pmatrix} = j_1 (1\ 0\ 0\ 0\ 1\ -1\ -1)^T.$$

Fig. 8.2 Loop current j_1

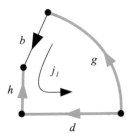

Fig. 8.3 Loop current j_2

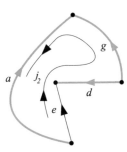

Therefore, the current vector i corresponds to the loop current j_1 flowing through the edges b, g, d, h and oriented as the cotree edge b, as shown in Fig. 8.2.

- For $i_e = j_2 \neq 0$ and $i_b = i_f = 0$, we have $i_C = j_2 (0\ 1\ 0)^T$ and

$$i_T = -j_2 x_2 = j_2 (-1\ 1\ -1\ 0)^T ; \quad i = j_2 (0\ 1\ 0\ -1\ 1\ -1\ 0)^T .$$

The loop current j_2 flows through the edges e, a, g, d and is oriented as the cotree edge e, as shown in Fig. 8.3. Note that the loop e, a, g, d is *not* a mesh.

- For $i_f = j_3 \neq 0$ and $i_b = i_e = 0$, we have $i_C = j_3 (0\ 0\ 1)^T$ and

$$i_T = -j_3 x_3 = j_3 (1\ -1\ 0\ 0)^T ; \quad i = j_3 (0\ 0\ 1\ 1\ -1\ 0\ 0)^T .$$

The loop current j_3 flows through the edges f, a, g and is oriented as the cotree edge f, as shown in Fig. 8.4.

Although the example refers to a planar graph for simplicity, the process used to obtain the loop currents is completely general and applies equally well to nonplanar graphs. Thus the loop currents represent a generalization of the mesh currents, which can be defined without ambiguity only in the case of a planar graph.[1] Choosing a different partition of the graph into a tree and cotree generates another set of

---

[1] There is no ambiguity provided that the graph is fixed and drawn without edge crossings.

Fig. 8.4 Loop current j_3

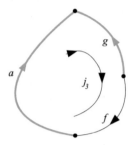

$L - N + 1$ (equivalent) loop currents, as you can easily verify. Each loop current, finally, is associated with the only cotree edge of the loop where it flows, and its orientation can always be chosen in agreement with that of such edge.

8.2 Tableau Method for Linear Time-Invariant Resistive Circuits

As shown in the first chapter, the most general method to find the complete solution (i.e., each descriptive current and voltage, collected in the vectors i and v, resp.) of a circuit is solving a set of independent topological equations (KVLs and KCLs), together with the set of descriptive equations related to all the components contained in the circuit.

In Chap. 2, it was shown that a possible way to obtain a set of independent topological equations is based on the fundamental matrices. In particular (see also Fig. 3.1),

$$Ai = 0_{N-1} \qquad (N - 1 \text{ independent KCLs in the unknowns } i) \qquad (8.2)$$

and

$$Bv = 0_{L-N+1} \qquad (L - N + 1 \text{ independent KVLs in the unknowns } v) \qquad (8.3)$$

where 0_K denotes the zero vector with K entries. Considering now the L descriptive equations (see Fig. 3.1), we know (see Chap. 3) that the most general way to express them for a memoryless circuit is the implicit form. Then for a linear time-invariant resistive circuit (i.e., a circuit containing only linear, time-invariant, memoryless components, and independent sources; see Sect. 3.3.3), the descriptive equations can be expressed as

$$H^v v + H^i i = \hat{u}(t) \qquad (L \text{ independent equations in the unknowns } v \text{ and } i) \qquad (8.4)$$

Fig. 8.5 Descriptive
variables for Case Study 1 in
Sect. 7.3

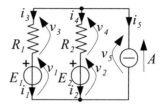

where H^v and H^i are square matrices (of size $L \times L$) whose entries have hybrid
physical dimensions and $\hat{u}(t)$ is an L-size vector whose (physically hybrid) entries
are either zeros or impressed currents/voltages. In particular, if the circuit contains
P independent sources, the vector $\hat{u}(t)$ will have $L - P$ null components.

For instance, for Case Study 1 in Sect. 7.3 with the complete set of unknowns
shown in Fig. 8.5, the descriptive equations for the five circuit components are:

$$
\begin{aligned}
v_1 &= E_1 \\
v_2 &= E_2 \\
-i_5 &= A \\
v_3 &= R_1 i_3 \\
v_4 &= R_2 i_4
\end{aligned}
\tag{8.5}
$$

which can be recast in the matrix form Eq. 8.4 as follows.

$$
\underbrace{\begin{pmatrix} 1&0&0&0&0 \\ 0&1&0&0&0 \\ 0&0&1&0&0 \\ 0&0&0&1&0 \\ 0&0&0&0&0 \end{pmatrix}}_{H^v}
\underbrace{\begin{pmatrix} v_1 \\ v_2 \\ v_3 \\ v_4 \\ v_5 \end{pmatrix}}_{v}
+
\underbrace{\begin{pmatrix} 0&0&0&0&0 \\ 0&0&0&0&0 \\ 0&0&-R_1&0&0 \\ 0&0&0&-R_2&0 \\ 0&0&0&0&-1 \end{pmatrix}}_{H^i}
\underbrace{\begin{pmatrix} i_1 \\ i_2 \\ i_3 \\ i_4 \\ i_5 \end{pmatrix}}_{i}
=
\underbrace{\begin{pmatrix} E_1 \\ E_2 \\ 0 \\ 0 \\ A \end{pmatrix}}_{\hat{u}(t)}
\tag{8.6}
$$

Then, the complete set of Eqs. 8.2–8.4 can be recast in a more compact form as
follows.

$$
\underbrace{\begin{array}{c} KCL \\ KVL \\ Descr \end{array}\begin{pmatrix} [0]_{(N-1)\times L} & A \\ B & [0]_{(L-N+1)\times L} \\ H^v & H^i \end{pmatrix}}_{Q}
\begin{pmatrix} v \\ i \end{pmatrix} = \begin{pmatrix} 0_L \\ \hat{u}(t) \end{pmatrix}
\tag{8.7}
$$

where $[0]_{a\times b}$ denotes the $a \times b$ zero matrix. The square (of size $2L \times 2L$) block
matrix Q in the l.h.s. of Eq. 8.7 is called the *tableau*.

If matrix Q is nonsingular (i.e., it can be inverted), the complete solution of the
circuit exists, is unique, and can be expressed as

$$\begin{pmatrix} v \\ i \end{pmatrix} = Q^{-1} \begin{pmatrix} 0_L \\ \hat{u}(t) \end{pmatrix} \tag{8.8}$$

Instead, if matrix Q is singular ($rank(Q) < 2L$), the circuit solution either does not exist or is not unique and in this case the circuit is commonly said to be *patholog-ical*. This means that the circuit contains at least one absurd connection of compo-nents, such as a loop of independent voltage sources, a cut-set of independent current sources, or a connection of components not compatible with the admitted basis or bases (e.g., an ideal transformer connected to both ports to two independent voltage sources or to two independent current sources).

As an example, taking $R_1 = R_2 = 0$ in the circuit of Fig. 8.5 (the corresponding graph edges remain unchanged), matrix Q becomes singular ($rank(Q) = 2L - 1$) and both spaces $\mathcal{N}(Q)$ and $\mathcal{N}(Q^T)$ are one-dimensional. (See Sect. 2.2.3.) More precisely, we have:

$$\mathbb{R}^{2L} = \mathcal{R}(Q) \oplus \mathcal{N}(Q^T); \quad \dim(\mathcal{N}(Q^T)) = \dim(\mathcal{N}(Q)) = 1$$

This means that the homogeneous equation $Qx = 0$ admits nontrivial solutions $x_0 \in \mathcal{N}(Q)$.

In the circuit, voltage sources E_1 and E_2 are now connected in parallel, therefore:

- For $E_1 \neq E_2$, the parallel connection of the voltage sources violates the KVL. In terms of matrix algebra, writing Eq. 8.7 as $Qx = b$ for ease of reference, it could be shown that the vector b has a component in the space $\mathcal{N}(Q^T)$, whereas any product Qx generates a vector in the (complementary) space $\mathcal{R}(Q)$; then the equation $Qx = b$ has no solutions.
- When $E_1 = E_2$ the parallel connection of the voltage sources does not imply a violation of the KVL and $b \in \mathcal{R}(Q)$, then equation $Qx = b$ can be solved. However, the currents i_1 and i_2 flowing through identical sources (or through short circuits, in the particular case when $E_1 = E_2 = 0$) can take any value such that $i_1 + i_2 = A$. In other terms, given two elements i_1 and $i_2 = A - i_1$ of the solution vector x, we can obtain a new solution by simply replacing i_1 with $i_1 + i_0$ and i_2 with $i_2 - i_0$ for any value of i_0, which means that the solution vector x is not unique; this amounts to saying that the addition of any vector $x_0 \in \mathcal{N}(Q)$ to a solution x generates a new solution $x + x_0$:

$$Q(x + x_0) = Qx + \underbrace{Qx_0}_{0} = Qx = b.$$

8.3   Superposition Theorem

In this section, we provide a proof of the superposition principle for linear time-invariant resistive circuits.

Theorem 8.1 (Superposition theorem) *Given a nonpathological linear time-invariant resistive circuit (see Sect. 3.3.3) containing P independent sources, its solution can be obtained as the sum of the solutions of P auxiliary circuits, each obtained by turning on one independent source at a time and turning off all the other independent sources.*

Proof The vector $\hat{u}(t)$ defined in the previous section can be recast as follows.

$$\hat{u}(t) = \begin{pmatrix} \hat{u}_1 \\ \hat{u}_2 \\ \vdots \\ \hat{u}_L \end{pmatrix} = \underbrace{\begin{pmatrix} \hat{u}_1 \\ 0 \\ \vdots \\ 0 \end{pmatrix}}_{\hat{u}^{(1)}} + \underbrace{\begin{pmatrix} \hat{u}_1 \\ \hat{u}_2 \\ \vdots \\ 0 \end{pmatrix}}_{\hat{u}^{(2)}} + \cdots + \underbrace{\begin{pmatrix} 0 \\ 0 \\ \vdots \\ \hat{u}_L \end{pmatrix}}_{\hat{u}^{(L)}} = \sum_{k=1}^{L} \hat{u}^{(k)} \tag{8.9}$$

We remark that only P of the L vectors $\hat{u}^{(k)}$ have one component different from 0. The other $L - P$ vectors have only null entries, thus the sum has only P significant terms. Because the names of the vectors $\hat{u}^{(k)}$ are completely arbitrary, we can assume without loss of generality that the significant vectors are the first P and that $\hat{u}(t) = \sum_{k=1}^{P} \hat{u}^{(k)}$.

Owing to the assumption of the nonpathological nature of the circuit, its solution can be expressed as

$$\begin{pmatrix} v \\ i \end{pmatrix} = Q^{-1} \begin{pmatrix} 0_L \\ \sum_{k=1}^{P} \hat{u}^{(k)} \end{pmatrix} \tag{8.10}$$

Inasmuch as Q does not depend on the index k, this is equivalent to

$$\begin{pmatrix} v \\ i \end{pmatrix} = \sum_{k=1}^{P} Q^{-1} \begin{pmatrix} 0_L \\ \hat{u}^{(k)} \end{pmatrix} \tag{8.11}$$

This means that the complete solution of the original circuit can be obtained by summing the solutions of P auxiliary circuits, obtained from the original one (the matrix Q is the same for all auxiliary circuits) by turning on only the kth independent source (corresponding to $\hat{u}^{(k)}$) and turning off all the other independent sources. □

8.4 Equivalent Representations of Memoryless Multiports

In this section, we generalize the concepts introduced in Sect. 3.4 for a memoryless two-terminal element to memoryless multiports. Also in this case, we aim to find a *macromodel* (i.e., an equivalent representation with the same descriptive equations but a simplified internal structure) of a portion of a linear time-invariant resistive circuit.

Fig. 8.6 Linear
time-invariant resistive
circuit represented as the
connection of an n-port $\aleph$
and n two-terminals

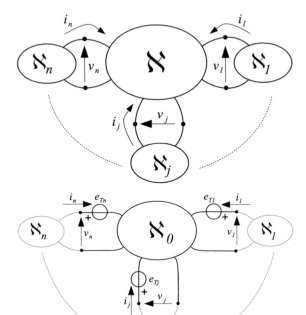

Fig. 8.7 Thévenin
equivalent representation (in
black) of the n-port $\aleph$ shown
in Fig. 8.6

We start by pointing out the presence of the n-port $\aleph$ within the circuit, as shown in Fig. 8.6. The picture is completed by n two-terminal (or one-port) macromodels $\aleph_j$ ($j = 1, \ldots, n$) representing the remaining parts of the original circuit.

General assumptions: Owing to the hypothesis of a linear time-invariant resistive circuit, we assume that $\aleph$ can contain only linear, time-invariant, memoryless components, and independent sources.

Temporary assumption: We also assume (only temporarily) that the circuit does not contain any controlled source. We remove this assumption later on.

There is also a further specific assumption about the basis admitted by $\aleph$, which is made in each of the following subsections. We remark that the proofs of the theorems provided therein are based on both the substitution principle (see Sect. 7.4) and the superposition theorem (see Sect. 8.3).

8.4.1 Thévenin Equivalent Representation of Memoryless Multiports

Theorem 8.2 (Thévenin equivalent representation of memoryless multiports) *Given an n-port $\aleph$ satisfying the general assumptions listed above and admitting the current basis, it can be represented equivalently by an n-port with the structure shown in Fig. 8.7.*

Fig. 8.8 Step 1 (substitution principle)

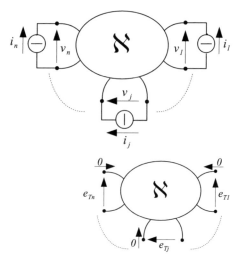

Fig. 8.9 Step 2a (first auxiliary circuit)

Proof The proof is based on two steps.

Step 1. Because $\aleph$ admits the current basis, we can substitute each two-terminal $\aleph_j$ ($j = 1, \ldots, n$) with an independent current source i_j, as shown in Fig. 8.8.

Step 2. Now we apply the superposition theorem to find the expressions of the port voltages v_j in terms of port currents and parameters of $\aleph$: this possibility is ensured by the fact that $\aleph$ admits the current basis. Due to the general assumptions about the composition of $\aleph$ and to the specific assumption about the admitted basis, its descriptive equations will have the following structure:

$$v = e_T + Ri \tag{8.12}$$

where e_T is a vector of n terms that are independent of the port currents i and R is an $n \times n$ matrix of real and constant coefficients.

In the following, the n-port $\aleph$ with all the internal sources turned off is denoted $\aleph_0$.

Step 2a. We obtain a first auxiliary circuit (Fig. 8.9) by turning off all the current sources i_j ($j = 1, \ldots, n$) and considering the contribution to v_j due to all the independent sources contained within $\aleph$. In other words, in this auxiliary circuit all ports are connected to open circuits (current sources turned off) and the corresponding open-circuit voltage on the jth port is the jth component of the vector e_T in Eq. 8.12. This term depends only on the impressed currents and voltages of the independent sources *internal* to $\aleph$ and on the parameters of the other components within $\aleph$. Of course, if the internal structure of $\aleph$ were known, we could consider one contribution for each independent source and the sum of all these terms would be e_{Tj}.

Step 2b. We now obtain other n auxiliary circuits: the kth auxiliary circuit (Fig. 8.10) is obtained by turning off all the sources internal to $\aleph$ (thus obtaining $\aleph_0$) and all the port current sources but i_k. The corresponding contribution to v_j (say $v_j^{(k)}$)

Fig. 8.10 Step 2b (*k*th auxiliary circuit)

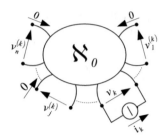

is obtained as follows.

$$v_j^{(k)} = v_j\big|_{\aleph=\aleph_0 \,\&\, i_j=0 \text{ for each } j\neq k} \tag{8.13}$$

Because $i_k \neq 0$, we can write that

$$v_j^{(k)} = \frac{v_j}{i_k}\bigg|_{\aleph=\aleph_0 \,\&\, i_j=0 \text{ for each } j\neq k} \qquad i_k = R_{jk}i_k \tag{8.14}$$

The term $R_{jk} = \dfrac{v_j}{i_k}\bigg|_{\aleph=\aleph_0 \,\&\, i_j=0 \text{ for each } j\neq k}$ has the physical dimension of a resistance and is an element of the matrix R in Eq. 8.12.

As an alternative (see Sect. 5.3.1 for $n = 2$), one can directly obtain the descriptive equations of the linear, time-invariant, and memoryless two-port $\aleph_0$ and identify from them the R entries.

Step 2c. Finally, we sum the contributions provided by the auxiliary circuits and obtain the expression of the port voltage v_j ($j = 1, \ldots, n$):

$$v_j = e_{Tj} + \sum_{k=1}^{n} v_j^{(k)} = e_{Tj} + \underbrace{\sum_{k=1}^{n} R_{jk}i_k}_{v_j'} \tag{8.15}$$

This equation allows us easily to obtain the Thévenin equivalent representation of $\aleph$: it tells us that the port voltage v_j is given by the sum of a voltage e_{Tj}, which is independent of the port currents and can be interpreted as an independent voltage source, and of a voltage v_j', which can be interpreted as the jth port voltage of the n-port $\aleph_0$, obtained by turning off all the internal independent sources of $\aleph$ and described by the resistance matrix R. This corresponds to the n-port shown in Fig. 8.7. $\square$

Remark 1. Step 2 in the proof requires the linearity assumption of $\aleph$, but nothing is said about the two-terminals $\aleph_j$ ($j = 1, \ldots, n$) representing the remaining parts of the original circuit. In fact, they could also be nonlinear, as often happens in electronic circuits.

Remark 2. Equation 8.12 is nothing but a generalization of Eq. 3.20, which is related to a two-terminal (particular case of n-port for $n = 1$) admitting the current

Fig. 8.11 Case Study 1

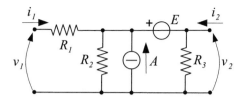

basis and made up of linear, time-invariant, and memoryless components and independent sources. The n-port $\aleph_0$ is compactly described by the resistance matrix R, whereas the contributions due to the internal sources are represented by the vector e_T.

Remark 3. If we remove the temporary assumption stated above and we admit the presence of controlled sources within $\aleph$, we have two possible cases:

- If the driving variable of a given controlled source belongs to $\aleph$ (i.e., it is either an inner variable or a port variable), from the $\aleph$ standpoint this source is in fact a controlled source and it is treated as such.
- On the contrary, if the driving variable is an inner variable of one of the two-terminals $\aleph_j$ ($j = 1, \ldots, n$), from the $\aleph$ standpoint this source is actually an independent source (there is no evidence of the presence of a driving variable, in $\aleph$) and it is treated as such; that is, it is turned off together with the independent sources during Step 2b.

Remark 4. Equation 8.12 holds for the Thévenin equivalent shown in Fig. 8.7. If the jth voltage source in Fig. 8.7 is connected upside down, there is just a change of sign in the corresponding component of the vector e_T, but we follow the same line of reasoning. This is apparent in Case Studies 2 and 3 below.

Remark 5. The proof provides a method to obtain the Thévenin equivalent representation of memoryless multiports, as shown in the next case studies.

Case Study 1

Find the Thévenin equivalent of the two-port shown in Fig. 8.11.

We can solve this problem in two different ways.

Way 1. We find the descriptive equations of the two-port in the form $v = e_T + Ri$ and we identify vector e_T and matrix R. This is left to the reader as an exercise.

Way 2. We analyze two simpler auxiliary circuits.

The first one (with port currents set to 0 through open circuits and port voltages equal to e_T) is shown in Fig. 8.12a and can be easily solved, thus finding $e_{T1} = \dfrac{R_2}{R_2 + R_3}(R_3 A + E)$ and $e_{T2} = \dfrac{R_3}{R_2 + R_3}(R_2 A - E)$.

The second auxiliary circuit (with independent sources turned off) is shown in Fig. 8.12b and provides $R = \dfrac{1}{R_2 + R_3}\begin{pmatrix} R_1 R_2 + R_1 R_3 + R_2 R_3 & R_2 R_3 \\ R_2 R_3 & R_2 R_3 \end{pmatrix}$.

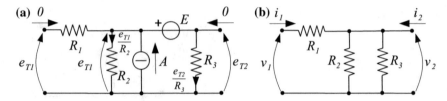

Fig. 8.12 Auxiliary circuits for Case Study 1

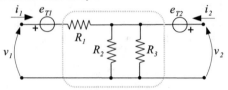

Fig. 8.13 Thévenin equivalent representation of Case Study 1

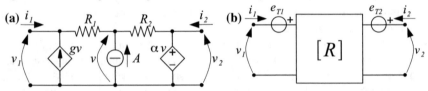

Fig. 8.14 Case Study 2

The Thévenin equivalent representation of the composite two-port shown in Fig. 8.11 is shown in Fig. 8.13, where the two-port $\aleph_0$ can be compactly represented by matrix R.

Of course, the second way is more suitable as far as the circuit becomes more complex.

Case Study 2

For the two-port shown in Fig. 8.14a, find the Thévenin equivalent shown in Fig. 8.14b.

We notice that in Fig. 8.14b one of the voltage sources is oriented differently than usual. This implies that in the first auxiliary circuit (with port currents set to 0) the port voltage v_1 is equal to $-e_{T1}$, whereas v_2 is equal to e_{T2}, as shown in Fig. 8.15a. This circuit can be easily solved, thus finding $e_{T1} = -\dfrac{1 + g R_1}{1 - \alpha - g R_2} R_2 A$ and $e_{T2} = \dfrac{\alpha R_2 A}{1 - \alpha - g R_2}$ (provided that the denominator is not null).

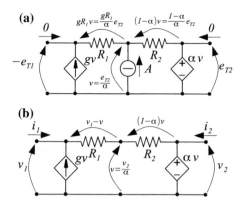

Fig. 8.15 Auxiliary circuits for Case Study 2

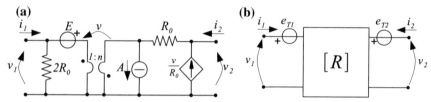

Fig. 8.16 Case Study 3

The second auxiliary circuit (with independent sources turned off) is shown in Fig. 8.15b and provides $R = \dfrac{1}{1 - \alpha - g R_2} \begin{pmatrix} R_2 + (1 - \alpha) R_1 & 0 \\ \alpha R_2 & 0 \end{pmatrix}$.

Case Study 3

For the composite two-port shown in Fig. 8.16a, with transformation ratio $n = 2$, find the Thévenin equivalent shown in Fig. 8.16b.

Also in this case, one of the voltage sources is oriented differently than usual. This implies that in the first auxiliary circuit (with port currents set to 0) the port voltage v_1 is equal to e_{T1}, whereas v_2 is equal to $-e_{T2}$, as shown in Fig. 8.17a. This circuit can be easily solved, thus finding $e_{T1} = \dfrac{4}{13} R_0 A - \dfrac{12}{13} E$ and $e_{T2} = -\dfrac{4}{13} R_0 A - \dfrac{1}{13} E$.

Fig. 8.17 Auxiliary circuits
for Case Study 3

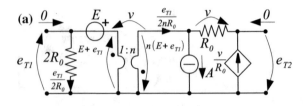

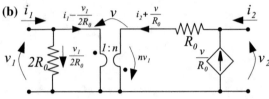

Fig. 8.18 Case Study 4: **a**
the circuit containing the
two-port ℵ and **b** the
Thévenin equivalent of ℵ

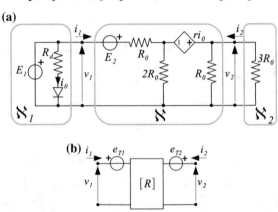

The second auxiliary circuit (with independent sources turned off) is shown
in Fig. 8.17b and provides $R = \dfrac{R_0}{13}\begin{pmatrix} 2 & -4 \\ 2 & 9 \end{pmatrix}$.

Case Study 4

*For the composite two-port ℵ, which is part of the circuit represented in
Fig. 8.18a, find the Thévenin equivalent shown in Fig. 8.18b.*

The circuit in Fig. 8.18a is nonlinear because of the presence of the diode in
ℵ₁. The two-port ℵ contains a CCVS whose driving current i_0 is just the diode's
current; in any case, because i_0 is external to ℵ, the descriptive equations of
the two-port can be obtained considering the CCVS as independent as well as
the voltage source E_2. The remaining components of ℵ are linear. Under the

Fig. 8.19 Auxiliary circuits for Case Study 4

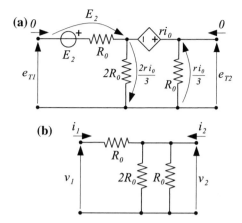

assumption that $\aleph$ admits the current basis, we can then obtain the parameters of its Thévenin equivalent shown in Fig. 8.18b.

Making reference to the auxiliary circuit shown in Fig. 8.19a, the open-circuit voltages e_{T1} and e_{T2} are: $e_{T1} = -E_2 - \dfrac{2r i_0}{3}$ and $e_{T2} = \dfrac{r i_0}{3}$. From the second auxiliary circuit (where the two sources are turned off) we easily obtain the resistance matrix:

$$R = R_0 \begin{pmatrix} \frac{5}{3} & \frac{2}{3} \\ \frac{2}{3} & \frac{2}{3} \end{pmatrix}.$$

8.4.2 Norton Equivalent Representation of Memoryless Multiports

Theorem 8.3 (Norton equivalent representation of memoryless multiports) *Given an n-port $\aleph$ satisfying the general assumptions listed at the beginning of Sect. 8.4 and admitting the voltage basis, it can be represented equivalently by an n-port with the structure shown in Fig. 8.20.*

Proof Also in this case, the proof is based on two steps.

Step 1. Because $\aleph$ admits the voltage basis, we can substitute each two-terminal $\aleph_j$ ($j = 1, \ldots, n$) with an independent voltage source v_j, as shown in Fig. 8.21.

Step 2. Now we apply the superposition theorem to find the expressions of the port currents i_j in terms of port voltages and parameters of $\aleph$: this possibility is ensured by the fact that $\aleph$ admits the voltage basis. Due to thegeneral assumptions

Fig. 8.20 Norton equivalent representation (in black) of the n-port $\aleph$ shown in Fig. 8.6

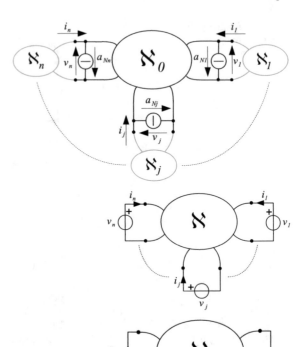

Fig. 8.21 Step 1 (substitution principle)

Fig. 8.22 Step 2a (first auxiliary circuit)

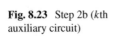

Fig. 8.23 Step 2b (kth auxiliary circuit)

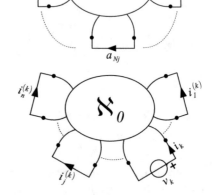

about the composition of $\aleph$ and to the specific assumption about the admitted basis, its descriptive equations have the following structure.

$$i = a_N + Gv \tag{8.16}$$

where a_N is a vector of n terms that are independent of the port voltages v and G is an $n \times n$ matrix of real and constant coefficients.

Step 2a. We obtain a first auxiliary circuit (Fig. 8.22) by turning off all the voltage sources v_j ($j = 1, \ldots, n$) and consider the contribution to i_j due to all the independent sources contained within $\aleph$. In other words, in this auxiliary circuit all ports are connected to short circuits (voltage sources turned off) and the corresponding short-circuit current on the jth port is the jth component of the vector a_N in Eq. 8.16. This term depends only on the impressed currents and voltages of the independent sources *internal* to $\aleph$ and on the parameters of the other components within $\aleph$. Of course, if the internal structure of $\aleph$ were known, we could consider one contribution for each independent source and the sum of all these terms would be a_{Nj}.

Step 2b. We now obtain other n auxiliary circuits: the kth auxiliary circuit (Fig. 8.23) is obtained by turning off all the sources internal to $\aleph$ (thus obtaining $\aleph_0$) and all the port voltage sources but v_k. The corresponding contribution to i_j (say $i_j^{(k)}$) is obtained as follows.

$$i_j^{(k)} = i_j \big|_{\aleph=\aleph_0 \ \& \ v_j=0 \text{ for each } j \neq k} \tag{8.17}$$

Because $v_k \neq 0$, we can write that

$$i_j^{(k)} = \frac{i_j}{v_k} \bigg|_{\aleph=\aleph_0 \ \& \ v_j=0 \text{ for each } j \neq k} v_k = G_{jk} v_k \tag{8.18}$$

The term $G_{jk} = \dfrac{i_j}{v_k} \bigg|_{\aleph=\aleph_0 \ \& \ v_j=0 \text{ for each } j \neq k}$ has the physical dimension of a conductance and is an element of the matrix G in Eq. 8.16.

As an alternative (see Sect. 5.3.2 for $n = 2$), one can directly obtain the descriptive equations of the linear, time-invariant, and memoryless two-port $\aleph_0$ and identify from them the G entries.

Step 2c. Finally, we sum the contributions provided by the auxiliary circuits and obtain the expression of the port current i_j ($j = 1, \ldots, n$):

$$i_j = a_{Nj} + \underbrace{\sum_{k=1}^{n} i_j^{(k)} = a_{Nj} + \underbrace{\sum_{k=1}^{n} G_{jk} v_k}_{i_j'}} \tag{8.19}$$

This equation allows us easily to obtain the Norton equivalent representation of $\aleph$: it tells us that the port current i_j is given by the sum of a current a_{Nj}, which is independent of the port voltages and can be interpreted as an independent current source, and of a current i_j', which can be interpreted as the jth port current of the n-port $\aleph_0$ obtained by turning off all the internal independent sources of $\aleph$ and described by the conductance matrix G. This corresponds to the n-port shown in Fig. 8.20. $\square$

The remarks listed in Sect. 8.4.1 hold also in this case, *mutatis mutandis*. Moreover, if $\aleph$ admits both current and voltage bases, $G = R^{-1}$.

Fig. 8.24 First auxiliary
circuit for Case Study 1

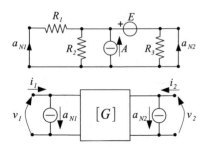

Fig. 8.25 Norton equivalent
representation of the
composite two-port shown in
Fig. 8.11

Case Study 1
Find the Norton equivalent of the composite two-port shown in Fig. 8.11.

Also in this case, we can solve this problem in two different ways.

Way 1. We find the descriptive equations of the two-port in the form $i = a_N + Gv$ and we identify vector a_N and matrix G. This is left to the reader as an exercise.

Way 2. We analyze two simpler auxiliary circuits.

The first one (with port voltages set to 0 through short circuits and port currents equal to a_N) is shown in Fig. 8.24 and can be easily solved, thus finding $a_{N1} = -\dfrac{E}{R_1}$ and $a_{N2} = E \dfrac{R_1 + R_2}{R_1 R_2} - A.$

The second auxiliary circuit $\aleph_0$ (with independent sources turned off) is the same as in Fig. 8.12b and provides $G = \dfrac{1}{R_1}\begin{pmatrix} 1 & -1 \\ -1 & 1 + \frac{R_1(R_2+R_3)}{R_2 R_3} \end{pmatrix}$. Taking the matrix R obtained in the Thévenin Case Study 1, you can easily check that the matrix product RG provides an identity matrix.

The Norton equivalent representation of the composite two-port (Fig. 8.11) is shown in Fig. 8.25, where the two-port $\aleph_0$ is compactly represented by matrix G.

Case Study 2
For the composite two-port shown in Fig. 8.26a, find the Norton equivalent shown in Fig. 8.26b.

The first auxiliary circuit (with port voltages set to 0 through short circuits) is shown in Fig. 8.27a. This circuit can be easily solved, thus finding $a_{N1} = -gE$ and $a_{N2} = A$.

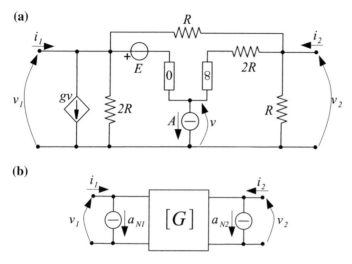

Fig. 8.26 Case Study 2

The second auxiliary circuit (with independent sources turned off) is shown in Fig. 8.27b and provides $G = \dfrac{1}{R}\begin{pmatrix} Rg + \frac{3}{2} & -1 \\ -1 & 2 \end{pmatrix}$.

Case Study 3

For the composite two-port shown in Fig. 8.28a, find the Norton equivalent shown in Fig. 8.28b.

In this case the original circuit is quite complex, but we can resort to another multiport description before obtaining the Norton equivalent. It is easy to show that the transmission matrices for two-ports A and B are $T_A = \begin{pmatrix} 0 & \frac{1}{g} \\ 0 & \frac{1}{Rg} \end{pmatrix}$ and $T_B = \begin{pmatrix} -\frac{1}{n} & 0 \\ 0 & -n \end{pmatrix}$, respectively. Because A and B are cascade-connected and the connection is admitted, the resulting two-port admits transmission matrix $T = T_A T_B = -\dfrac{n}{Rg}\begin{pmatrix} 0 & R \\ 0 & 1 \end{pmatrix}$.

In the Norton equivalent the current sources are oriented differently than usual. This implies that in the first auxiliary circuit the short-circuit port current i_1 is equal to $-a_{N1}$, whereas i_2 is equal to $-a_{N2}$, as shown in Fig. 8.29a. This circuit can be easily solved, thus finding $a_{N1} = \dfrac{E}{R}$ and $a_{N2} = A + \dfrac{gE}{n}$.

(a)

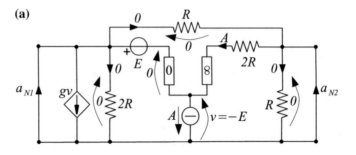

(b)

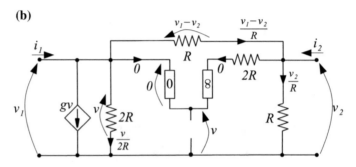

Fig. 8.27 Auxiliary circuits for Case Study 2

(a)

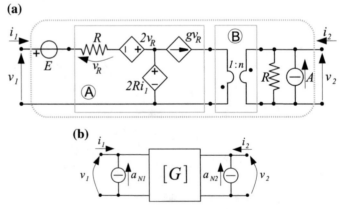

(b)

Fig. 8.28 Case Study 3

The second auxiliary circuit $\aleph_0$ (with independent sources turned off) is shown in Fig. 8.29b and provides $G = \dfrac{1}{R}\begin{pmatrix} 1 & 0 \\ \frac{Rg}{n} & 1 \end{pmatrix}$.

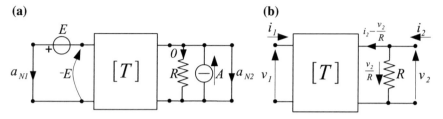

(a)   **(b)**

Fig. 8.29   Auxiliary circuits for Case Study 3

Fig. 8.30   Case Study 4.
Norton equivalent (**a**) for the
two-port ℵ of Fig. 8.18a; first
auxiliary circuit (**b**)

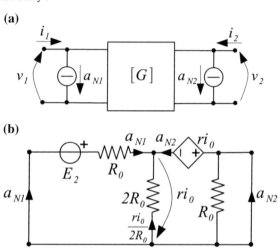

(a)

(b)

Case Study 4

*For the composite two-port ℵ shown in Fig. 8.18a, find the Norton equivalent
shown in Fig. 8.30a.*

Following the discussion developed for Case Study 4 of Sect. 8.4.1, the
CCVS must be treated as an independent voltage source because its driving
current i_0 is external to the two-port. From the auxiliary circuit of Fig. 8.30b
we easily obtain $a_{N1} = \dfrac{E_2 + ri_0}{R_0}$ and $a_{N2} = \dfrac{-3ri_0}{2R_0} - \dfrac{E_2}{R_0}$. The matrix G is
found from the circuit of Fig. 8.19b:

$$G = \frac{1}{2R_0}\begin{pmatrix} 2 & -2 \\ -2 & 5 \end{pmatrix}.$$

Obviously, this matrix is the inverse of the R matrix found for the Thévenin
equivalent.

(a)   **(b)**

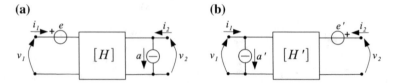

Fig. 8.31 Hybrid equivalent representations for memoryless two-ports

8.4.3 Hybrid Equivalent Representations of Memoryless Multiports

Theorem 8.4 (Hybrid equivalent representation of memoryless multiports) *Given an n-port ℵ satisfying the general assumptions listed at the beginning of Sect. 8.4 and admitting a mixed basis, it can be represented equivalently by an n-port ℵ$_0$ (obtained by ℵ turning off all the internal independent sources and described by a hybrid matrix) with either a voltage source in series (as for the Thévenin equivalent) or a current source in parallel (as for the Norton equivalent) to each port, according to the admitted mixed basis.*

The proof is completely similar to the Thévenin and Norton cases and is not reported here.

For the case of memoryless two-ports ($n = 2$), the two possible hybrid equivalent representations are shown in Fig. 8.31.

8.5 Summarizing Comments

Most of the results proposed in this chapter rely on the linearity assumption. Despite the fundamental laws of physics are often assumed to be linear, the real world is usually nonlinear: then the linearity assumption can have validity limits, more or less hard, as pointed out at the beginning of this book (Sect. 1.1). Whenever possible, it is useful to resort to this assumption in order to exploit analysis tools that are available only for linear circuits and systems.[2] In this perspective, linearity can be either an intrinsic property of a given physical system or, more commonly, a working assumption with validity limits that must be known. As an example for the first case, the electric field vector E originated by a set of charges in the free space can be obtained by summing the field contributions of the individual charges, thanks to the linearity of the electric field equations. As to the second case, a physical resistor maintains its linear behavior until the current through it does not cause overheating, which alters the resistance value (and makes it dependent on the current magnitude) or destroys it (as occurs for an electric fuse). Similarly, the variation of a spring length depends linearly on the applied force (Hooke's law) as long as this force

---

[2]In his Lectures on Physics, while introducing linear systems, the Nobel laureate Richard Feynman writes: "[···] linear systems are so important [···] because we can solve them" [1].

does not exceed a certain limit. Beyond this limit the change in length no longer follows a linear law. Furthermore, for large variations with respect to the length of the undeformed spring, the elasticity assumption for the spring is no longer valid and permanent deformations may occur.

The main advantage of linear systems is that they can be broken down into parts (e.g., the auxiliary circuits we introduced with the superposition principle) that can be solved separately: the overall solution is a linear combination of the partial solutions obtained in this way. This is the essence of the superposition principle, which allows greatly simplifying complex problems. In other words, a linear system is equal to the sum of these parts. Actually, it has been known for many centuries that often, "The whole is more than the sum of its parts," which is the abridged version of the sentence, "The totality is not, as it were, a mere heap, but the whole is something besides the parts." [2]. In fact, most of everyday life is nonlinear and the superposition principle fails due to interferences, competitions, and cooperations between "parts".

There are many examples coming from the real world, for instance, the emergence of flocking, swarming, and schooling in groups of agents (e.g., birds, fish, penguins, ants, bees, bacteria) that follow three basic heuristic rules: cohesion (attempting to stay close to nearby agents), collision avoidance, and velocity matching [3–5]. Despite the fact that each agent tries to follow these very simple rules and has only local interactions with the other agents, the overall behaviors turn out to fulfill different main optimization criteria such as time, energy, and predation risk in different situations [6]. Other familiar examples are concerned with [7–9]:

- Consciousness, which is an emergent property that engages incredibly high numbers of neurons (corresponding to agents) and synapses (corresponding to local connections)
- Magnetism, which emerges from the collective behavior of millions of spins
- Large-scale electrical blackouts, resulting from a cascading series of failures in electrical power grids
- Spread of information and disease, strongly affected by structure and dynamics of animal social networks
- Extinction of living groups, resulting from the interactions between the elements of food chains
- Spread of computer viruses, depending on both the Internet structure and the behaviors of computer users
- Traffic jams, emerging from car drivers interacting on urban streets

In all cases, the "whole" depends on both the topology (structure of the hardware infrastructure and the interactions between agents/parts) and the agent/part individual way of functioning/behaving. This is an aspect common to linear and nonlinear systems. But for linear systems (including circuits), the individual behaviors are simple and there cannot be surprising behaviors emerging from their connections. Moreover, in the circuits considered in this book, most components are time-invariant, which in turn greatly simplifies the analysis, excluding the possibility of changes in both network topology and component behaviors.

(a)   **(b)**

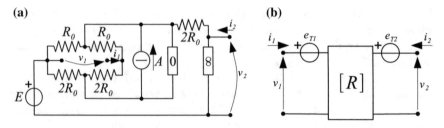

Fig. 8.32 Problem 8.1

The difficulties increase more and more as far as we introduce in a network (of any kind) nonlinear agents/parts and plasticity in the interactions (i.e., time variance in the network topology). We had a taste of this in Sect. 3.2.5, when we analyzed simple circuits containing a diode: in order to solve these circuits analytically, we had to resort to piecewise-linear approximations of the diode DP characteristic. But doing this simply means that we replace the originally smooth nonlinear DP characteristic with linear approximations that are valid under some limits (voltage/current range); outside a given range, we change the linear approximation.

This is the simplest way to maintain both many advantages of the linearity and the nonlinear nature of the circuit. This is also the working principle of many circuit simulators, which simulate nonlinear circuits iteratively, bringing them back to an equivalent linear circuit at each iteration. It may take many iterations before the calculations converge to a solution.

We remark that the increase in complexity in the nonlinear case can be twigged in memoryless circuits, but it will become more and more evident when considering dynamical circuits and systems in Volume 2.

8.6 Problems

8.1 For the two-port shown in Fig. 8.32a, find the Thévenin equivalent representation shown in Fig. 8.32b. *Hint*: When analyzing the second auxiliary circuit (to find the matrix R), either add an auxiliary unknown for one of the resistors in the lattice structure (at your choice), or exploit the information provided by the nullor (in fact, the two resistors with resistance $2R_0$ on the lower side of the lattice behave as if they were connected in series).

8.2 Determine the descriptive equations of the two-port shown in Fig. 8.33a. Then, find, if admitted, the Thévenin and Norton equivalent representations shown in Fig. 8.33b and c, respectively.

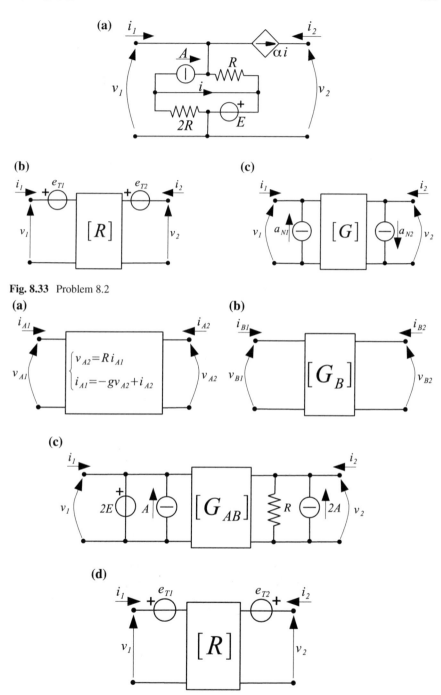

Fig. 8.33 Problem 8.2

Fig. 8.34 Problem 8.3

Fig. 8.35 Problem 8.4

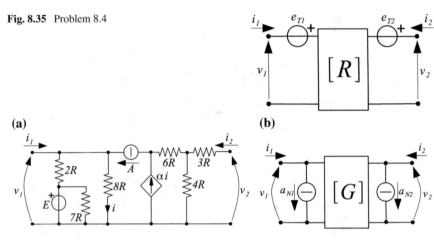

(a)

(b)

Fig. 8.36 Problem 8.5

8.3 Given the two two-ports A and B shown in Fig. 8.34 (panels a and b, resp.),
where $G_B = \frac{1}{R} \begin{pmatrix} 1 & 2 \\ 0 & 0 \end{pmatrix}$:

1. Find the bases and matrices admitted by A.
2. Is the cascade connection of A and B admitted? And of A and a nullor? And of B and an ideal transformer?
3. Find the reciprocity condition, if any, for A.
4. Find the conductance matrix G_{AB} of the two-port obtained by connecting A and B in parallel.
5. For the two-port shown in Fig. 8.34c (where G_{AB} is the conductance matrix found at the previous step), find the Thévenin equivalent representation shown in Fig. 8.34d.

8.4 For a two-port described by the descriptive equations $v_1 = \alpha v_2 + RA - E$ and $i_2 = g v_1 + 2A$ (with $\alpha, g \neq 0$):

1. Find the Thévenin equivalent representation shown in Fig. 8.35.
2. Determine if a Norton equivalent representation would also be admitted.
3. By considering the two-port with independent sources turned off, determine its properties and admitted bases and matrices.

8.5 For the two-port shown in Fig. 8.36a, find the Norton equivalent representation shown in Fig. 8.36b.

8.6 For the two-port shown in Fig. 8.37a, where $\alpha \geq 0$ and $\beta = \frac{3}{2}$, find the Norton equivalent representation shown in Fig. 8.37b.

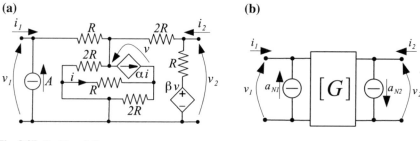

Fig. 8.37 Problem 8.6

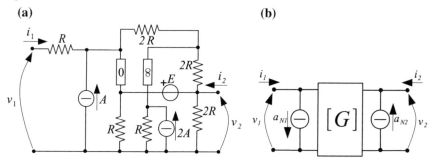

Fig. 8.38 Problem 8.7

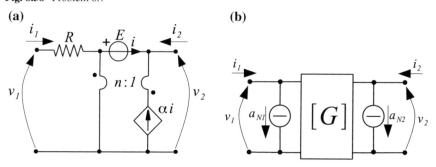

Fig. 8.39 Problem 8.8

8.7 For the two-port shown in Fig. 8.38a, find the Norton equivalent representation shown in Fig. 8.38b.

8.8 For the two-port shown in Fig. 8.39a, with $n \neq \alpha$, find the Norton equivalent representation shown in Fig. 8.39b.

References

1. Feynman RP, Kreyszig E, Leighton RB, Sands M (1977) The Feynman lectures on physics. Addison-Wesley, Reading

2. Aristotle: Metaphysics, Book H, 1045a:8–10
3. Okubo A (1986) Dynamical aspects of animal grouping: swarms, schools, flocks, and herds. Adv Biophys 22:1–94
4. Vicsek T (2001) A question of scale. Nature 411:421–421
5. Olfati-Saber R (2006) Flocking for multi-agent dynamic systems: algorithms and theory. IEEE Trans Autom Control 51:401–420
6. Alerstam T (2011) Optimal bird migration revisited. J Ornithol 152:5–23
7. Strogatz SH (2001) Exploring complex networks. Nature 410:268–276
8. Albert R, Barabási AL (2002) Statistical mechanics of complex networks. Rev Mod Phys 74:47–97
9. Boccaletti S, Latora V, Moreno Y, Chavez M, Hwang DU (2006) Complex networks: structure and dynamics. Phys Rep 424:175–308

Appendix
Synoptic Tables

A.1 Metric System Prefixes

Table A.1 summarizes the most used prefixes of the metric system.

Table A.1 Metric system prefixes

Prefix	Symbol	Value	Colloquial
femto	f	10^{-15}	quadrillionth
pico	p	10^{-12}	trillionth
nano	n	10^{-9}	billionth
micro	μ	10^{-6}	millionth
milli	m	10^{-3}	thousandth
centi	c	10^{-2}	hundredth
deci	d	10^{-1}	tenth
/	/	10^{0}	one
deka	da	10^{1}	ten
hecto	h	10^{2}	hundred
kilo	k	10^{3}	thousand
mega	M	10^{6}	million
giga	G	10^{9}	billion
tera	T	10^{12}	trillion
peta	P	10^{15}	quadrillion

© Springer International Publishing AG 2018
M. Parodi and M. Storace, *Linear and Nonlinear Circuits:*
Basic & Advanced Concepts, Lecture Notes in Electrical Engineering 441,
DOI 10.1007/978-3-319-61234-8

Table A.2 Properties of the main memoryless components

Component	Linear	Time-invariant	Bases	Energetic behav.	Recipr.	Symmetr.
Resistor	Y	Y	v, i	Passive	Y	/
Voltage source	N	N	i	Active	N	/
Current source	N	N	v	Active	N	/
Diode (Shockley)	N	Y	v, i	Passive	N	/
CCCS	Y	Y	(i_1, v_2)	Active	N	N
CCVS	Y	Y	(i_1, i_2)	Active	N	N
VCVS	Y	Y	(v_1, i_2)	Active	N	N
VCCS	Y	Y	(v_1, v_2)	Active	N	N
Nullor	N	Y	/	Active	N	N
Ideal transformer	Y	Y	(v_1, i_2) (i_1, v_2)	Nonenergic	Y	N (if $n \neq 1$)
Gyrator	Y	Y	(i_1, i_2) (v_1, v_2)	Nonenergic	N	N

A.2 Properties of the Main Memoryless Components

Table A.2 summarizes the fundamental properties of the most relevant memoryless components.

A.3 Reciprocity and Symmetry Conditions for Linear, Time-Invariant, and Memoryless Two-Ports

Table A.3 summarizes the reciprocity and symmetry conditions for linear, time-invariant, and memoryless two-ports described by at least one matrix.

Table A.3 Reciprocity and symmetry conditions for linear, time-invariant, and memoryless two-ports described by at least one matrix

Matrix	Reciprocity condition	Symmetry condition	Reciprocity and symmetry condition
R	$R_{12} = R_{21}$	$R_{12} = R_{21}$ $R_{11} = R_{22}$	$R_{12} = R_{21}$ $R_{11} = R_{22}$
H	$H_{12} = -H_{21}$	$H_{11} = \dfrac{H_{11}}{det(H)}$ $H_{12} = -\dfrac{H_{21}}{det(H)}$ $H_{21} = -\dfrac{H_{12}}{det(H)}$ $H_{22} = \dfrac{H_{22}}{det(H)}$	$H_{12} = -H_{21}$ $det(H) = 1$
T	$det(T) = 1$	$T_{11} = \dfrac{T_{22}}{det(T)}$ $T_{12} = \dfrac{T_{12}}{det(T)}$ $T_{21} = \dfrac{T_{21}}{det(T)}$ $T_{22} = \dfrac{T_{11}}{det(T)}$	$det(T) = 1$ $T_{11} = T_{22}$

Solutions

Problems of Chap. 1

Solutions of Problems of Chap. 1.

1.1 $i_x = 4$ A.

1.2 $i_1 = -1$ A, $i_2 = 1$ A, $i_3 = 5$ A, $i_4 = -5$ A.

1.3 $v_x = 5$ V.

1.4 $v_1 = 1$ V, $v_2 = -4$ V, $v_3 = 5$ V, $v_4 = -1$ V.

1.5 $p(t) = -i_A v_3 - i_B v_2 - i_C (v_1 + v_3)$.

Problems of Chap. 2

Solutions of Problems of Chap. 2.

2.1 For the choice of tree with edges a, b, h, and m, the fundamental cut-set matrix

$$\text{is } A = \begin{array}{c} \\ a \\ b \\ h \\ m \end{array} \begin{array}{cccccc} c & d & e & f & g & l \end{array} \quad \begin{array}{cccc} a & b & h & m \end{array}$$

$$\text{is } A = \begin{array}{c} a \\ b \\ h \\ m \end{array} \left(\begin{array}{cccccc|cccc} 0 & 0 & 0 & -1 & -1 & 1 & 1 & 0 & 0 & 0 \\ 0 & -1 & 1 & 1 & 1 & -1 & 0 & 1 & 0 & 0 \\ -1 & -1 & 1 & 1 & 1 & 0 & 0 & 0 & 1 & 0 \\ 1 & 1 & 0 & 0 & -1 & 0 & 0 & 0 & 0 & 1 \end{array} \right).$$

The corresponding fundamental loop matrix B can be easily obtained owing to the relationships between A and B.

2.2 Following the hint (due to the fact that there are no components between the lowest connections/dots), the graph corresponding to the circuit has $N = 8$ nodes and $L = 12$ edges. Then, we need $N - 1 = 7$ independent KCLs and $L - N + 1 = 5$ independent KVLs.

© Springer International Publishing AG 2018
M. Parodi and M. Storace, *Linear and Nonlinear Circuits:*
Basic & Advanced Concepts, Lecture Notes in Electrical Engineering 441,
DOI 10.1007/978-3-319-61234-8

2.3 We have 5 unknown currents and 3 independent KCLs, then 2 independent currents, say i and i_3. If we measure i_3, i cannot be determined. The graph of Fig. 2.26b is the same as for the circuit of Fig. 2.26a, then, one can apply Tellegen's theorem, thus finding: $\hat{v}_1 i_1 + \hat{v}_2 i_2 + \hat{v}_3 i_3 + \hat{v}_4 i + \hat{v}_5 i_4 = 0$. Taking into account the KCLs ($i_2 = i_3$, $i_1 = -i - i_3$, and $i_4 = i$), one finds $(\hat{v}_4 + \hat{v}_5 - \hat{v}_1)i = (\hat{v}_1 - \hat{v}_2 - \hat{v}_3)i_3$. But KVLs imply that the terms between brackets are identically zero, thus it is not possible to determine i knowing i_3.

2.4 The graph has $N = 6$ nodes and $L = 8$ edges. Thus it has $L - N + 1 = 3$ fundamental loops and $N - 1 = 5$ fundamental cut-sets and as many tree edges.

Problems of Chap. 3

Solutions of Problems of Chap. 3.

3.1 $v_{AC} = (R_1 - R_2)i_C + R_1 i_A$; $v_{CB} = (R - R_1 + R_2)i_C + (R - R_1)i_A$.

3.2

Set	Linear	Time-invariant	Memory	Physical dimension
(a)	Y	Y	Y	$\beta, \gamma : [\Omega]; \delta : [\Omega^{-1}s]$
(b)	N	Y	Y	$\alpha : [\Omega^{-2}V^{-1}]; \gamma : [A^{-1}]; \delta : [\Omega s]$
(c)	Y	Y	N	$\beta, \gamma, \sigma : [\Omega]$
(d)	N	Y	N	$\beta : [V]; I_0 : [A\, rad^{-1}]; \gamma : [\Omega^{-1}]; \sigma : [/]$

3.3 (a)

1. $v = (R_1 + R_2)i - R_1 A - E$.
2. Both bases admitted.
3. Active (it would be passive only in the particular case with $R_1 A + E = 0$, assuming that the resistance values are positive).
4. $R_{TH} = R_{NR} = R_1 + R_2$; $e_{TH} = -R_1 A - E$; $a_{NR} = \dfrac{R_1 A + E}{R_1 + R_2}$.

(b) The four resistors are in parallel, thus equivalent to a single resistor with resistance $R = \dfrac{R_1 R_2 R_3 R_4}{R_1 R_2 (R_3 + R_4) + R_3 R_4 (R_1 + R_2)}$. Therefore

1. $v = Ri$.
2. Both bases admitted.
3. Passive.
4. $R_{TH} = R_{NR} = R$; $e_{TH} = 0$V; $a_{NR} = 0$A.

3.4 (a) Active; (b) active.

3.5 (a) $R_{TH} = R_{NR} = R$; $e_{TH} = E - RA$; $a_{NR} = A - E/R$.

(b) $R_{TH} = R_{NR} = \dfrac{R_3(R_1 + R_2)}{R_1 + R_2 + R_3}$; $e_{TH} = \dfrac{R_3(E + R_1 A)}{R_1 + R_2 + R_3}$; $a_{NR} = -\dfrac{e_{TH}}{R_{TH}} = $

$-\dfrac{E + R_1 A}{R_1 + R_2}$.

3.6 $i = \dfrac{R_2 E_1 + R_1 E_2}{R_1 R_2 + R_2 R_3 + R_1 R_3} = 1$ mA.

3.7

1. $i = \dfrac{R_2 A - E}{R_1 + R_2}$.

2. $v = \dfrac{R_2 \left[(R_1 + R_3)A + E \right] + R_1 R_3 A}{R_1 + R_2}$.

3. $p_A = -\dfrac{R_1 R_2 + R_2 R_3 + R_1 R_3}{R_1 + R_2} A^2 - \dfrac{R_2 A E}{R_1 + R_2}$.

4. $p_E = -\dfrac{E^2}{R_1 + R_2} + \dfrac{R_2 A E}{R_1 + R_2}$.

5. No. From an energetic point of view, this is due to the fact that the power absorbed by resistors is always positive: if the powers absorbed by the sources were also positive, we would have an absurd energy balance. From a purely mathematical point of view, you can easily check that p_A can be positive only if the product AE is negative, whereas p_E can be positive only if the product AE is positive.

3.8 $R_1 = R_2$.

3.9 The only right answer is the first one.

3.10 $a_{NR} = -\dfrac{e}{R}$; $R_{NR} = \dfrac{R}{1 + Rg_2}$.

3.11

1. $i = 0$ for $v \in (-1, 1)$; $v = 1$ V for $i > 0$; $v = -1$ V for $i < 0$.
2. $p = -2$ W.

3.12 $v = \dfrac{G_1 E + A}{G_1 + G_2 + G_3} \left(= \dfrac{E + R_1 A}{R_1 R_2 + R_2 R_3 + R_1 R_3} R_2 R_3 \right) = 5$V.

3.13 $e_{TH} = 3RA = 7.5$ V; $R_{TH} = 4R = 100\,\Omega$.

Problems of Chap. 5

Solutions of Problems of Chap. 5.

5.1 $R = \begin{pmatrix} R_1 + (\beta + 1)R_3 & R_3 \\ (\beta + 1)R_3 + \beta R_2 & R_2 + R_3 \end{pmatrix}.$

The other matrices (provided that they are admitted) can be expressed in terms of the resistance matrix (this is a general property):

$$G = \begin{pmatrix} \dfrac{R_{22}}{det(R)} & -\dfrac{R_{12}}{det(R)} \\[2mm] -\dfrac{R_{21}}{det(R)} & \dfrac{R_{11}}{det(R)} \end{pmatrix} \quad H = \begin{pmatrix} \dfrac{det(R)}{R_{22}} & \dfrac{R_{12}}{R_{22}} \\[2mm] -\dfrac{R_{21}}{R_{22}} & \dfrac{1}{R_{22}} \end{pmatrix} \quad T = \begin{pmatrix} \dfrac{R_{11}}{R_{21}} & \dfrac{det(R)}{R_{21}} \\[2mm] \dfrac{1}{R_{21}} & \dfrac{R_{22}}{R_{21}} \end{pmatrix}$$

5.2 $e_{TH} = 0,\ R_{TH} = \dfrac{3R^2 + 4R R_0}{4R + 4R_0},\ R_0 = \dfrac{\sqrt{3}}{2}R$

5.3 (a) $G = \dfrac{1}{R_1 R_2 + R_1 R_3 + R_2 R_3} \begin{pmatrix} R_2 + R_3 & -R_3 \\[2mm] -R_3 & \dfrac{R_1 R_2 + R_1 R_3 + R_2 R_3 + R_1 R_4 + R_3 R_4}{R_4} \end{pmatrix}.$

The other matrices (provided that they are admitted) can be expressed in terms of the conductance matrix (this is a general property):

$$R = \begin{pmatrix} \dfrac{G_{22}}{det(G)} & -\dfrac{G_{12}}{det(G)} \\[2mm] -\dfrac{G_{21}}{det(G)} & \dfrac{G_{11}}{det(G)} \end{pmatrix} \quad H = \begin{pmatrix} \dfrac{1}{G_{11}} & -\dfrac{G_{12}}{G_{11}} \\[2mm] \dfrac{G_{21}}{G_{11}} & \dfrac{det(G)}{G_{11}} \end{pmatrix} \quad T = \begin{pmatrix} -\dfrac{G_{22}}{G_{21}} & -\dfrac{1}{G_{21}} \\[2mm] -\dfrac{det(G)}{G_{21}} & -\dfrac{G_{11}}{G_{21}} \end{pmatrix}.$$

The two-port is reciprocal and can be symmetrical (if $G_{11} = G_{22}$).

(b) $R = \begin{pmatrix} R_1 + R_2 & R_2 \\ R_2 & R_2 + R_3 \end{pmatrix}.$

The other matrices (provided they are admitted) can be expressed in terms of the resistance matrix (see solution to Problem 5.1). The two-port is reciprocal and can be symmetrical (if $R_1 = R_3$).

(c) $H = \begin{pmatrix} R_1 & \alpha \\ \beta & \dfrac{1}{R_2} \end{pmatrix}.$

The other matrices (provided they are admitted) can be expressed in terms of the first hybrid matrix (this is a general property):

$$R = \begin{pmatrix} \dfrac{det(H)}{H_{22}} & \dfrac{H_{12}}{H_{22}} \\[2mm] -\dfrac{H_{21}}{H_{22}} & \dfrac{1}{H_{22}} \end{pmatrix} \quad G = \begin{pmatrix} \dfrac{1}{H_{11}} & -\dfrac{H_{12}}{H_{11}} \\[2mm] \dfrac{H_{21}}{H_{11}} & \dfrac{det(H)}{H_{11}} \end{pmatrix} \quad T = \begin{pmatrix} -\dfrac{det(H)}{H_{21}} & -\dfrac{H_{11}}{H_{21}} \\[2mm] -\dfrac{H_{22}}{H_{21}} & -\dfrac{1}{H_{21}} \end{pmatrix}.$$

The two-port can be reciprocal (if $\alpha = -\beta$) and can be symmetrical (if $\alpha = -\beta$ and $R_1/R_2 = 1 - \beta^2$; see Sect. 5.5.2 and Table A.3).

(d) $T = \begin{pmatrix} \alpha & 0 \\ \dfrac{(\alpha - 1)R_2 - R_1}{\beta R_1 R_2} & -\dfrac{1}{\beta} \end{pmatrix}$, provided that $\beta \neq 0$.

The other matrices (provided they are admitted) can be expressed in terms of the forward transmission matrix (this is a general property):

$$R = \begin{pmatrix} \dfrac{T_{11}}{T_{21}} & \dfrac{det(T)}{T_{21}} \\ \dfrac{1}{T_{21}} & \dfrac{T_{22}}{T_{21}} \end{pmatrix} \quad G = \begin{pmatrix} \dfrac{T_{22}}{T_{12}} & -\dfrac{det(T)}{T_{12}} \\ -\dfrac{1}{T_{12}} & \dfrac{T_{11}}{T_{12}} \end{pmatrix} \quad H = \begin{pmatrix} \dfrac{T_{12}}{T_{22}} & \dfrac{det(T)}{T_{22}} \\ -\dfrac{1}{T_{22}} & \dfrac{T_{21}}{T_{22}} \end{pmatrix}.$$

The two-port can be reciprocal (if $\alpha = -\beta$) and can be symmetrical (if $\beta = 1$ and $\alpha = -1$ or if $\beta = -1$ and $\alpha = 1$).

5.4 (a) If $g \neq 1/R$, all bases are admitted. If $g = 1/R$, only bases (v_1, v_2) and (v_1, i_2) are admitted. $T = \begin{pmatrix} 2 & R \\ \dfrac{1 - Rg}{R} & 1 - Rg \end{pmatrix}$.

(b) Admitted bases: (i_1, i_2) and (v_1, i_2). T is not admitted.

5.5 (a) $i = \dfrac{R_2 + R_3}{R_3(R_1 + R_2)}e$.

(b) $v_o = -\dfrac{R_2}{R_1}e_i$ (inverting voltage amplifier).

(c) $v_o = \dfrac{R_1 + R_2}{R_1}e_i$ (noninverting voltage amplifier).

5.6

1. $p = \dfrac{E_1^2}{2R} = 2$ mW.
2. $v_1 = -3E_1 = -6$ V.
3. $v_3 = 6E_1 + \frac{4}{3}E_2 = 16$ V.

5.7 $I = \dfrac{R_2 A - nE}{n(R_1 + R_2)} = 6$ mA.

5.8

1. $i = -\dfrac{3}{5}\dfrac{E}{R} = -3$ mA.
2. $p = -6\dfrac{E^2}{R} = -15$ mW.

5.9 $I = \dfrac{\beta R_2 - r}{(1 + \beta)(R_2 + r)} \dfrac{E}{R_1} = 10$ mA, by assuming $\beta \neq -1$ and $r \neq -R_2$.

5.10 $\dfrac{p_u}{p_E} = \dfrac{R_u}{n^2 R_s + R_u}$ (< 1).

5.11

1. $i_\infty = E \dfrac{(1 + g R_4) R_2}{g R_1 R_2 (R_3 + R_4) + R_3 (1 + g R_4)(R_1 + R_2)} = 200$ mA;

2. $v_\infty = -\dfrac{R_3 + R_4}{1 + g R_4} i_\infty = -2$ V.

3. $p = -R_3 i_\infty^2 = -400$ mW.

5.12

1. $i_\infty = \dfrac{3}{2} A - \dfrac{2E}{R}$.

2. $v_\infty = 5E - 4RA$.

3. $p = \dfrac{2E^2}{R} - \dfrac{3}{2} AE$.

4. $p = -AE$.

5.13

1. $p = E \left(A + \dfrac{nE}{2R} \right) \dfrac{2n}{n^2 + 3} = 12$ mW.

2. $p = 0$. (It is an ideal power transferitor by definition.)

5.14 $e_{TH} = \dfrac{(R_1 + R_2) R_3 A - \alpha R_3 (R_2 A + E)}{(1 + \alpha) R_1 + R_2}$ and $R_{TH} = R_3$, provided that $\alpha \neq$

$- \left(1 + \frac{R_2}{R_1} \right)$.

5.15

1. (i_{A1}, i_{B1}), (v_{A1}, v_{B1}), (v_{A1}, i_{B1}).

2. (i_{A2}, i_{B2}), (i_{A2}, v_{B2}), (v_{A2}, i_{B2}).

3. $T = \begin{pmatrix} \dfrac{1}{2} + Rg & -R \\ \dfrac{g}{2} & 0 \end{pmatrix}$.

4. $v_1 = \dfrac{A_1}{g}(1 + 2Rg) - RA_2$.

5. $v_2 = \dfrac{2A_1}{g}$.

5.16

1. $v = \dfrac{3}{8}E = 1.5$ V.

2. $p = -\dfrac{v^2}{2R} = -45$ mW.

5.17 $e_{TH} = -\dfrac{RA}{n(GR + \alpha)}$; $R_{TH} = \dfrac{1}{n^2}\dfrac{\beta R + r}{GR + \alpha}$.

$a_{NR} = \dfrac{e_{TH}}{R_{TH}} = -\dfrac{nRA}{\beta R + r}$; $R_{NR} = R_{TH}$.

5.18 $v_A = -\dfrac{3}{5}(E + 3RA)$.

5.19

1. $v_1 = -\beta R i_2$ and $v_2 = R(1 - \beta)i_2$.

2. Admitted bases: (i_1, i_2) and (i_1, v_2); admitted matrices: R, H, T' (by assuming $\beta \neq 0$ and $\beta \neq 1$).

3. $[R] = R \begin{pmatrix} 0 & -\beta \\ 0 & 1 - \beta \end{pmatrix}$.

4. The two-port cannot be symmetric if we assume (as usual) $R > 0$.

5.20

1. $v = \dfrac{E}{3} + \dfrac{2}{9}RA = 400$ mV.

2. $p = -\dfrac{2}{9}A\left(E + \dfrac{2}{3}RA\right) = -8$ mW.

5.21

1. $R_{eq} = \dfrac{5}{3}R$.

2. $e_{TH} = -\dfrac{E}{4}$; $R_{TH} = \dfrac{5}{8}R$.

5.22

1. $i_E = \dfrac{E}{R} - \dfrac{2}{3}A = -160$ mA.

2. $p = RAi_E = -2.4$ W.

5.23

1. Admitted bases: (i_1, i_2), (v_1, i_2) and (i_1, v_2); $T_1 = -\begin{pmatrix} \dfrac{1}{2} & 0 \\ \dfrac{5}{3R} & \alpha \end{pmatrix}$; reciprocity

   condition: $\alpha = 2$.

2. $T_2 = \begin{pmatrix} 0 & -R \\ -\dfrac{1}{2R} & 0 \end{pmatrix}$.

3. $T = T_1 T_2 = \begin{pmatrix} 0 & \dfrac{R}{2} \\ \dfrac{\alpha}{2R} & \dfrac{5}{3} \end{pmatrix}$.

4. No.

5. Yes, provided that the overall two-port admits the current basis.

5.24 $a_{NR} = -\dfrac{E}{R}$; $R_{NR} = \dfrac{R}{2}$.

5.25 $p = \dfrac{6}{5}EA - \dfrac{4}{25}\dfrac{E^2}{R}$.

5.26

1. $v_1 = (\beta + 1)v_2 + 2R(\beta + 1)i_2$ and $v_2 = Ri_1$.

2. (i_1, i_2), (v_1, i_2), and (v_1, v_2).

3. All but H.

4. $\beta = -\dfrac{1}{2}$.

5. $T = \begin{pmatrix} \beta + 1 & -2R(\beta + 1) \\ \dfrac{1}{R} & 0 \end{pmatrix}$.

6. $e_{TH} = -\dfrac{nE}{\beta + 1} - 2RA = -2nE - 2RA$; $R_{TH} = -R$.

5.27

1. $a_{NR} = \dfrac{E}{(\alpha - 1)R_1}$; $R_{NR} = \dfrac{(1 - \alpha)R_1 R_2}{(1 - \beta)R_1 + R_2}$.

2. $p = -\dfrac{E^2}{2R}$.

5.28

1. $i = \dfrac{e_1 - e_2}{R_0}$.

2. It can be used as a voltage-to-current converter.

5.29 $i = \dfrac{R_4}{R_1} \dfrac{e}{\left[R_4 - R_0 \left(\dfrac{R_3}{R_2} - \dfrac{R_4}{R_1} \right) \right]}$. Note: if $\dfrac{R_3}{R_2} = \dfrac{R_4}{R_1}$, then $i = \dfrac{e}{R_1}$ independently of R_0; that is, the two-port connected to the voltage source (port 1) and to the resistor R_0 (port 2) is a VCCS!

Problems of Chap. 7

Solutions of Problems of Chap. 7.

7.1 $i = \dfrac{R_2 E_1 + R_1 E_2}{R_1 R_2 + R_2 R_3 + R_1 R_3} = 1$ mA.

7.2

1. $i = \dfrac{R_2 A - E}{R_1 + R_2}$.

2. $v = \dfrac{R_2[(R_1 + R_3)A + E] + R_1 R_3 A}{R_1 + R_2}$.

3. $p_A = -\dfrac{R_1 R_2 + R_2 R_3 + R_1 R_3}{R_1 + R_2} A^2 - \dfrac{R_2 A E}{R_1 + R_2}$.

4. $p_E = -\dfrac{E^2}{R_1 + R_2} + \dfrac{R_2 A E}{R_1 + R_2}$.

7.3 $v = \dfrac{G_1 E + A}{G_1 + G_2 + G_3} = 5$ V $\left(= \dfrac{E + R_1 A}{R_1 R_2 + R_2 R_3 + R_1 R_3} R_2 R_3 \right)$.

7.4

1. $p = \dfrac{E_1^2}{2R} = 2$ mW.

2. $v_1 = -3E_1 = -6$ V.

3. $v_3 = 6E_1 + \frac{4}{3}E_2 = 16$ V.

7.5 $I = \dfrac{R_2 A - nE}{n(R_1 + R_2)} = 6$ mA.

7.6

1. $i_\infty = \dfrac{3}{2}A - \dfrac{2E}{R}$.

2. $v_\infty = 5E - 4RA$.

3. $p = 2\dfrac{E^2}{R} - \dfrac{3}{2}AE$.

4. $p = -AE$.

7.7

1. $p = E\left(A + \dfrac{nE}{2R}\right)\dfrac{2n}{n^2 + 3} = 12$ mW.

2. $p = 0$ (it is an ideal power transferitor by definition).

7.8 $v_A = -\dfrac{3}{5}(E + 3RA)$.

7.9

1. $v = \dfrac{E}{3} + \dfrac{2}{9}RA = 400 \text{ mV}.$

2. $p = -\dfrac{2}{9}A\left(E + \dfrac{2}{3}RA\right) = -8 \text{ mW}.$

7.10

1. $i_E = \dfrac{E}{R} - \dfrac{2}{3}A = -160 \text{ mA}.$

2. $p = RAi_E = -2.4 \text{ W}.$

7.11 $\quad p = \dfrac{6}{5}EA - \dfrac{4}{25}\dfrac{E^2}{R}.$

7.12 $\quad p = -\dfrac{E^2}{2R}.$

7.13 $\quad i = \dfrac{e_1 - e_2}{R_0}.$

7.14

1. $i \approx 90 \text{ mA}.$

2. $p \approx 3.93 \text{ W}.$

7.15

1. $i_1 \approx -124 \text{ mA}.$

2. $i_2 \approx -1.12 \text{ A}.$

3. $p \approx 1.48 \text{ W}.$

7.16

1. $p \approx -308 \text{ mW}.$

2. $i \approx 262 \text{ mA}.$

3. $v \approx 2.92 \text{ V}.$

7.17

1. $i \approx -381 \text{ mA}.$

2. $p \approx 533 \text{ mW}.$

Problems of Chap. 8

Solutions of Problems of Chap. 8.

8.1

1. $e_{T1} = -2R_0A$, $e_{T2} = 8R_0A - E$.

2. $R = \frac{2}{3}R_0 \begin{pmatrix} 2 & 0 \\ 1 & 0 \end{pmatrix}$.

8.2

1. Descriptive equations: $i_1 = \frac{v_1}{R} - \frac{\alpha+2}{2R}E - (\alpha + 1)A$; $i_2 = \alpha(A + \frac{E}{2R})$.

2. The two-port does not admit the current basis and admits the voltage basis; then only its Norton equivalent representation exists.

3. $a_{N1} = A + \frac{E}{R} + \alpha\frac{2RA+E}{2R}$, $a_{N2} = \alpha\left(A + \frac{E}{2R}\right)$.

4. $G = \begin{pmatrix} \frac{1}{R} & 0 \\ 0 & 0 \end{pmatrix}$.

8.3

1. Admitted bases: (v_{A1}, v_{A2}) and (v_{A1}, i_{A2}); admitted matrices: G, H', T'.

2. Both A and B do not admit matrix T, thus they cannot be connected in cascade with anything.

3. We can find matrix $G_A = \frac{1}{R}\begin{pmatrix} 0 & 1 \\ 0 & 1 + Rg \end{pmatrix}$. Then (because $G_{12} \neq G_{21}$), A cannot be reciprocal.

4. $G_{AB} = G_A + G_B = \frac{1}{R}\begin{pmatrix} 1 & 3 \\ 0 & 1 + Rg \end{pmatrix}$.

5. $e_{T1} = 2E$, $e_{T2} = \frac{2RA}{2 + Rg}$, $[R] = \begin{pmatrix} 0 & 0 \\ 0 & \frac{R}{2+Rg} \end{pmatrix}$.

8.4

1. $e_{T1} = \frac{2A}{g}$, $e_{T2} = \frac{1}{\alpha}\left(E - RA - \frac{2A}{g}\right)$, $[R] = \frac{1}{g}\begin{pmatrix} 0 & 1 \\ 0 & \frac{1}{\alpha} \end{pmatrix}$.

2. A Norton equivalent is not admitted, because the two-port does not admit the voltage basis. Notice that the Thévenin matrix $[R]$ is singular.

3. Properties: linear, time-invariant, memoryless, active, nonreciprocal, nonsymmetrical; admitted bases: (i_1, i_2), and (i_1, v_2); admitted matrices: R, H, T'.

8.5 $a_{N1} = -A - \dfrac{E}{2R}$, $a_{N2} = \dfrac{4}{7}A$, $G = \dfrac{1}{R}\begin{pmatrix} \frac{5}{8} & 0 \\ -\frac{\alpha}{14} & \frac{1}{7} \end{pmatrix}$. Notice that the two-port

$\aleph_0$ with internal sources turned off would be reciprocal for $\alpha = 0$: you can state that either by using the reciprocity theorem (the CCCS is the only nonreciprocal element of $\aleph_0$ and disappears for $\alpha = 0$) or by inspection of matrix G.

8.6 $a_{N1} = A$, $a_{N2} = 0$, $G = \dfrac{1}{R}\begin{pmatrix} \frac{1}{2} & -\frac{1}{4} \\ -1 & 1 \end{pmatrix}$.

8.7 $a_{N1} = -\dfrac{E}{R}$, $a_{N2} = -\dfrac{E}{2R} - A$, $G = \dfrac{1}{R}\begin{pmatrix} 1 & -1 \\ 1 & \frac{1}{2} \end{pmatrix}$.

8.8 $a_{N1} = -\dfrac{E}{R}$, $a_{N2} = \dfrac{n(\alpha+1)}{(n-\alpha)}\dfrac{E}{R}$, $G = \dfrac{1}{R}\begin{pmatrix} 1 & -1 \\ -\frac{n(\alpha+1)}{(n-\alpha)} & \frac{n(\alpha+1)}{(n-\alpha)} \end{pmatrix}$.

Index

A
Adynamic, *see* memoryless
Ammeter, 7

B
Basis of definition, 66
Bias, 169

C
Cascade connection, 148
Colored edge theorem, 174
Component, 6
Conductance matrix, 127
Controlled source, 115
    CCCS, 115
    CCVS, 117
    nonlinear, 121
    VCCS, 120
    VCVS, 119
Cotree, 30
Current, 7
Current divider
    resistive, 88
Cut-set, 14, 28
    nodal, 29

D
Descriptive equations, 50, 230
    BJT, 63
    CCCS, 116
    CCVS, 118
    current source, 53
    diode, 59

    resistor, 51
    VCCS, 120
    VCVS, 119
    voltage source, 53
Descriptive variables, 6
    n-port, 114
    n-terminal, 16
    two-terminal, 8
Directionality, 138
Driving-point characteristic, 61

E
Edge, 23
Energetic behavior, 68
    active, 68
    dissipative, 68
    locally active, 171
    nonenergic, 68, 166
    passive, 68, 166
    strictly active, 68
Energy, 11

F
Full-wave bridge rectifier, 177
Fundamental cut-set matrix, 31
Fundamental loop matrix, 34

G
Graetz bridge, 176
Graph, 23
    connected, 27
    hinged, 28
    isomorphic, 24

© Springer International Publishing AG 2018
M. Parodi and M. Storace, *Linear and Nonlinear Circuits:*
Basic & Advanced Concepts, Lecture Notes in Electrical Engineering 441,
DOI 10.1007/978-3-319-61234-8

planar, 24
   star, 25
Gyrator, 145

H
Hybrid matrices, 129

I
Ideal power transferitors, 139
Ideal transformer, 140
Iterative resistance, 216

K
Kirchhoff's laws, *see* asotopological equa-
        tions12, 50, 102
   KCL, 15
   KVL, 13
   matrix formulation, 30

L
Ladder structure
   resistive, 214
Lattice (bridge) structure, 211
Linear, 64, 248
Loop, 12, 28
   inner loop, 14
   mesh, 14, 28
   outer loop, 14

M
Maxwell's minimum heat theorem, 109
Memoryless, 65, 256
Mesh analysis, 191
   generalized, 227
   modified, 196
   pure, 192
Millmann's rule, 205

N
Network, 165
Nodal analysis, 185
   modified, 189
   pure, 185
Node, 23
   order, 24
No-gain theorems, 178, 180
Norton equivalent
   n-port, 241

two-terminal, 76
n-port, 113
Nullor, 121

O
Ohm's law, 51
Operational amplifier, 123

P
Parallel
   two-port, 151
   two-terminal, 82
Parallel-series connection, 151
Path, 27
Π structure, 208
Port, 113
Potential function, 93
   cocontent, 96, 100, 104
   content, 94, 100, 103
Power
   absorbed, 9, 19, 114
Problem solutions
   Chap. 1, 259
   Chap. 2, 259
   Chap. 3, 260
   Chap. 5, 261
   Chap. 7, 267
   Chap. 8, 268

R
Reciprocity, 69
   theorem, 167
   two-port, 132
Resistance matrix, 126
Resistive, *see* memoryless

S
Series
   two-port (series-series), 150
   two-terminal, 80
Series-parallel connection, 151
Small signal, 169
Standard choice, 19, 114
   two-terminal, 9
Stationary point, 103
Subgraph, 27
Substitution
   principle, 204
   theorem, 204
Superposition

principle, 199
theorem, 233
Symmetry, 136

T
Tableau, 231
Tellegen's theorem, 43, 56, 165
Thévenin equivalent
    n-port, 234
    two-terminal, 73
Time-invariant, 64
Topological equations, *see* aso Kirchhoff's
    laws50, 230
Transmission matrix, 130

Tree, 29
T structure, 207

V
Variational principle, 102
Voltage, 7
Voltage divider
    resistive, 87
Voltmeter, 8

W
Wattmeter, 10

Printed in the United States
By Bookmasters